PARASITES AND PARASITIC DISEASES

NEMATODES

COMPARATIVE GENOMICS, DISEASE MANAGEMENT AND ECOLOGICAL IMPORTANCE

PARASITES AND PARASITIC DISEASES

Additional books in this series can be found on Nova's website under the Series tab.

Additional E-books in this series can be found on Nova's website under the E-book tab.

PARASITES AND PARASITIC DISEASES

NEMATODES

COMPARATIVE GENOMICS, DISEASE MANAGEMENT AND ECOLOGICAL IMPORTANCE

LEE M. DAVIS
EDITOR

nova publishers
New York

NOTICE TO THE READER

Library of Congress Cataloging-in-Publication Data

LCCN: 2013955541

ISBN: 978-1-62948-764-9

Published by Nova Science Publishers, Inc. † New York

CONTENTS

PREFACE

Nematodes are a highly assorted group of invertebrates occupying many terrestrial and aquatic habitats. Recent research has focused on the examination of nematode life history and biological traits to reflect the key functional attributes and infer the effects of environmental variables and human activities. In this book, the authors discuss the interaction between lectins and nematodes, biodiversity and distribution of nematodes belonging to the families *Longidoridae* and *Trichodoridae* in different habitats, analytical method for chemical characterization of nematodes, and the use of plants as natural sources of nematicides.

Chapter 1 – Apart from using species diversity and abundance data for analysis of nematode community in response to environmental conditions, recent research has focused on the examination of nematode life history and biological traits to reflect the key functional attributes and infer the effects of environmental variables and human activities. Based on a range of biological traits, including feeding type, tail shape, adult body length, body shape and life history, the response of nematode community upon abatement of sewage pollution in a subtropical harbour environment was investigated in a spatio-temporal scale. Results showed that the benthic environment at the inner-harbour sampling sites was still polluted, according to the composition of deposit and epigrowth feeders, body shape and length, and the combined biological trait analysis. However, data from the colonizer-persister scale were not consistent with the above observations. The present findings suggested that biological trait analysis can be complementary to community composition analysis in deciphering the response of nematode communities from the cessation of sewage pollution, while the use of a single trait or index should be cautioned.

Chapter 2 – The Nematoda phylum is a numerous and diversified group of metazoans. According to estimates, it includes between 100,000 and 1,000,000 free-living (terrestrial or aquatic) and parasitic species that are found parasitizing animals (vertebrates and invertebrates) as well as plants, with beneficial or harmful effects. Entomoparasitic nematodes (EPNs) living symbiotically with bacteria are used in biological control strategies against insects in agriculture. Sedentary phytonematodes like Meloidogyne incognita are among some of the main agricultural pathogens, responsible for annual losses accounting to US$125 billion a year. Currently, control strategies against these parasites rely on the use of chemical vermicides that may not be as efficient as they once were. Apart from this, these chemicals are harmful to the environment, extending their effects to other species, even mammals. In this sense, new and more viable control strategies that are less environmentally noxious have become increasingly important. In this sense, more recently lectins have been successfully used as a new model in the study of nematodes. Lectins are non-immune hemagglutinating proteins or glycoproteins constitutively expressed in seeds, leaves, barks, roots, tubers and fruit of several plant species, though these compounds may also be found in all kinds of organisms. They are biodegradable, and therefore do not build up in the soil or in organisms that feed on species controlled with lectins. This chapter describes how lectins may Interact with nematodes, reviews assays of the use of plant lectins to control phytonematodes, and how lectins are used by EPN.

Chapter 3 – This article presents information related to the biodiversity and distribution of nematodes belonging to the families *Longidoridae* and *Trichodoridae* in different habitats in the countries belonging to the CIS (Commonwealth of Independent States, formerly known as the USSR).The investigations have been conducted in 13 republics of the former USSR, and in 83 regions of the Russian Federation. The studies revealed 40 species of *Longidoridae,* including 18 species belonging to the genus *Longidorus,* 3 belonging to the genus *Paralongidorus*, and 19 belonging to *Xiphinema*. The list of species belonging to the family *Trichodoridae* is represented by 12 species, including 9 species belonging to the genus *Trichodorus,* and 3 belonging to *Paratrichodorus.* Most of these nematodes are characteristic for the palearctic fauna. The combination of field and laboratory studies allowed us to assess the importance of various agroecological factors (e.g. host plants, geographical distribution, location, climatic conditions, edaphic characteristics/soil properties, etc.) for population abundance and the occurrence of particular species.

Chapter 4 – Nematodes have a key importance in most aquatic and terrestrial ecosystems, independently of their lifestyle. Moreover, these worms play an important role as bioindicators in the investigation of organic and inorganic chemical pollutants as well. The free-living nematodes are used in biomonitoring due to their specific sensitivity and tolerance patterns towards changes in the environment, for example metal pollution of their habitat. For this reason, broadening our knowledge on their metal uptake and release characteristics will increase our understanding on the various roles they play in contaminated habitats by providing explanations to their stress responses. Free-living nematodes are microscopic in size and have a relatively short life-cycle. Due to these reasons, and because of their quite complex life processes, they are used as biological model organisms. In order to follow the uptake, distribution, detoxification and biomineralization processes in these animals, special analytical techniques are necessary due to the microscopic size (e. g. Ø=50 μm) of most free-living nematode species. In this review, the sample preparation procedures for elemental analysis of individual nematodes are demonstrated. Methods which preserve the original structure and enable the investigation of distribution and speciation of micro- and trace elements in small biological samples are discussed. A brief introduction will be provided concerning the microanalytical techniques used for the determination of trace and microelement content of nematodes, such as atomic absorption spectrometry, inductively coupled plasma atomic emission and mass spectrometry, total reflection X-ray fluorescence spectrometry. Furthermore, analytical methods applied for the investigation of elemental distribution (scanning and transmission electron microscopy, synchrotron radiation induced X-ray fluorescence spectrometry) and chemical speciation (X-ray absorption near edge structure, extended X-ray absorption fine structure) are reviewed. In addition to the principle of these instrumental techniques, their analytical figures of merit are also highlighted. A few characteristic analytical results and observations will be demonstrated e.g. element uptake routes (cuticular or intestinal), metal binding by different organs, biotransformation of heavy metals and nanoparticles, biogenic mineral generation within the body of nematodes, as well as the passage of polluting elements. The information obtained by these analytical techniques will increase our knowledge on biological effects of polluting heavy metals and may provide explanation on the stress responses of nematodes, these biological indicators and regulator organisms of very high ecological importance.

Chapter 5 – Plant parasitic nematodes can cause serious damages to numerous economically important agricultural crops worldwide. Control of these pests is based on pesticides and on the use of resistant cultivars. Although chemicals are effective in the plant parasitic nematode control, the recent European legislation has deeply revised and restricted their use on agricultural crops focusing the attention on environmental safety and human and animal health. Moreover, resistant cultivars are available only for a limited number of crops and plant parasitic nematode species. During the last decade research on low environmental impact alternatives to nematicides has received a strong impulse and a wide range of options including agronomic strategies, physical methods, bio-pesticides, and biocidal plants haveas been considered. Focusing on plants derived natural plant products known as botanic pesticides can represent useful tools to minimize soil nematode population density under the tolerance limits of the different crops to specific phytoparasitic nematodes. These botanic pesticides are generally non-persistent in field conditions as they are readily transformed by light, oxygen and microorganisms into non-toxic products without residues in the environment. Many botanic pesticides derived from wild and cultivated plants have been shown to possess nematicidal properties against several plant-parasitic nematodes. The aim of this review is to report the effects of botanic pesticides derived from various wild, cultivated and medicinal plants, including essential oils, root and tissues extracts, on plant parasitic nematodes in the view of sustainable agriculture.

In: Nematodes
Editor: Lee M. Davis

ISBN: 978-1-62948-764-9

Chapter 1

BIOLOGICAL TRAIT ANALYSIS OF NEMATODE COMMUNITY UPON ANTHROPOGENIC DISTURBANCES IN A SUBTROPICAL HARBOUR ENVIRONMENT

Wenzhe Xu[1], Xiaoshou Liu[1,2], Zhinan Zhang[2], Siu Gin Cheung[1,3] and Paul K. S. Shin[1,3*]

[1]Department of Biology and Chemistry, City University of Hong Kong, Hong Kong SAR, China
[2]College of Marine Sciences, Ocean University of China, Qingdao, China
[3]State Key Laboratory in Marine Pollution, City University of Hong Kong, Hong Kong SAR, China

ABSTRACT

Apart from using species diversity and abundance data for analysis of nematode community in response to environmental conditions, recent research has focused on the examination of nematode life history and

* Corresponding author: Tel 852-34427720, Fax 852-34420522, Email: bhpshin@cityu.edu.hk.

biological traits to reflect the key functional attributes and infer the effects of environmental variables and human activities. Based on a range of biological traits, including feeding type, tail shape, adult body length, body shape and life history, the response of nematode community upon abatement of sewage pollution in a subtropical harbour environment was investigated in a spatio-temporal scale. Results showed that the benthic environment at the inner-harbour sampling sites was still polluted, according to the composition of deposit and epigrowth feeders, body shape and length, and the combined biological trait analysis. However, data from the colonizer-persister scale were not consistent with the above observations. The present findings suggested that biological trait analysis can be complementary to community composition analysis in deciphering the response of nematode communities from the cessation of sewage pollution, while the use of a single trait or index should be cautioned.

INTRODUCTION

Free-living marine nematodes have been suggested as possible pollution indicators as they possess characteristics such as a short life-span and high diversity which make them potentially useful in ecological monitoring (Heip 1980; Heip et al. 1985). They are extremely abundant and diverse, often numbering millions per square meter in sediments, and occur in more habitats than any other metazoan group (Heip et al. 1985) which makes it easy to collect enough specimens in small volume of samples. Furthermore, they play an important role in the benthic ecosystem and food chain (Gee 1989; Coull 1999). In most studies, species diversity and community structure from species abundance data were used to analyze a community's response to environmental conditions (Heip et al. 1985). However, the application of the trophic structure (Wieser 1953) and/or the maturity index (Bongers 1990) have been demonstrated as other approaches to reflect the effects of anthropogenic activities on the nematode communities (Bongers et al. 1991; Essink and Keidel 1998; Gyedu-Ababio et al. 1999; Boyd et al. 2000; Guo et al. 2002). There are also studies using a functional group approach to examine the changes of the nematode community under environmental perturbations based on the natural-history information in nematode ecology (Thistle and Sherman 1985). More recently, attempts have been made to embrace a range of biological traits, derived from the attributes of nematode feeding habits, life histories, etc., to reflect some key functional attributes and infer the effects of environmental variables and human activities (Schratzberger et al. 2007).

Biological trait analysis is a method recently proposed for describing ecological functioning (such as matter cycling, energy flow and modification on habitats) of freshwater or marine benthic assemblages (Statzner et al. 1994; Bremner et al. 2006). It uses a series of life history, morphological and behavioural characteristics of the species present in benthic communities to indicate aspects of their ecological functioning. It then incorporates the information on species' distributions and the biological characteristics they exhibit, to produce a summary of the biological trait composition of benthic communities. The approach provides a link between species, environments and ecosystem processes, and is potentially useful for the investigation of anthropogenic impacts on ecological functioning (Bremner et al. 2006). In addition, this method has advantages over other approaches in describing ecological functioning, as it incorporates information on both the relative abundance/biomass of the species present in assemblages and their biological characteristics into one analysis, and it can encompass information on a large range of traits which are potentially important for ecological functioning. It is a complement to community structure analysis in which general descriptions of benthic diversity and community patterns are analyzed based on taxonomic species composition (Schratzberger et al. 2007).

From the last century, several morphological features of free-living marine nematodes thought to be related to important ecological functions have been proposed. Wieser (1953) found that the buccal structures of marine nematode genera could be used to classify them into different feeding guilds and this classification has been applied in many studies (Heip et al. 1985; Giere 1993). Nematode tail shapes were also proposed as a feature for functional grouping since tails could be important in locomotion and reproduction (Thistle and Sherman 1985). Organism size has always been treated as an important ecological feature as some physiological functions (such as metabolic rates, tolerance to chemical stress, ability to move or migrate and vulnerability to predation) clearly depend on it (Peters 1983; Soetaert et al. 2002). Nematode body length and width are also described as important functional attributes of the nematode species and genera (Soetaert et al. 2002; Vanaverbeke et al. 2003, 2004). Jensen (1987) suggested that nematode body length and shape could reflect the response of nematodes to environmental stresses (e.g., hypoxia and sulphide rich in sediments), in addition to adaptations to the physical constraints of interstitial space. According to their life history strategies, Bongers (1990) developed a grouping of nematode genera to assess the condition of terrestrial and freshwater habitats. Partly owing to a lack of empirical support for the classification of some marine genera and the absence or rarity of extreme colonizers and persisters of nematodes in most marine habitats, this classification, relating the more *r*-selected 'colonizers' to the more *K*-selected 'persisters', has been applied less frequently to marine nematodes (Bongers et al. 1991; Fraschetti et al. 2006). A very limited functional diversity using a smaller number of single functional groups, which reduces the generally high species diversity, will bring potential risks in underestimating the true functional complexity of nematode communities (Thistle et al. 1995). In order to overcome this shortcoming, Schraztberger et al. (2007) combined a range of biological traits of the nematode species and genera describing buccal morphology, tail shape, body size, body shape and life history strategy to identify patterns in the functional diversity of nematodes from the southwestern North Sea. They found that the functional composition of nematodes was strongly linked to median particle diameter and silt content of the sediment and water depth.

Nematodes are permanent members of the benthos and therefore are unable to physically escape from the effects of pollution in the benthic environment. In particular, nematodes offer a potential means of assessing sediment quality (Platt et al. 1984; Heip et al. 1985). Apart from changes of species diversity and community structure, the functioning group composition of the nematode community was also altered under pollution (Bongers et al. 1991; Gyedu-Ababio et al. 1999; Liu et al. 2008). The present study examined the response of nematode communities, based on analyses of biological traits, to organic pollution in a subtropical harbour.

MATERIALS AND METHODS

Field Sampling and Treatment of Samples

The field study was undertaken in Victoria Harbour, Hong Kong (22°20'N; 114°10'E), which has been receiving preliminary treated sewage of 1.7 million m^3 daily from some 3 million people for many years until 2001. Since then, with proper sewage treatment, the water quality of the habrour has been improved (EPD 2004). However, the sediment still contains relatively high organic content which may continuously exert an adverse impact on the benthic communities (Liu et al. 2011). In the present study, samples were collected at five locations across the harbour from east to west (Figure 1). Sites CW, SWH and CB were located within the harbour proper, whereas sites TLC and TY were outside the main harhour waters. Field sampling at these locations was conducted once in the wet (summer) and dry (winter) season, since in subtropical Hong Kong, there are marked differences in the water regime, e.g., temperature, noted between these seasons (Morton and Morton 1983). A Global Position System was used to fix the position of these sampling sites. At each sampling site, six sediment samples were collected with a 0.1 m^2 van Veen grab. Three sediment samples were used for nematode analysis, with samples being taken using a cut-off syringe (D=3 cm × L=8 cm) from each replicate grab sample. All the animal samples were fixed with 4% buffered formalin immediately on board and followed by staining with 1% Rose Bengal. Surface sediment for particle size and chemical analyses, including total organic carbon (TOC), total Kjeldahl nitrogen (TKN) and total phosphorus (TP), were collected from another three grab samples. All sediment samples for chemical analyses were stored in an ice box on board and transferred to the laboratory for further processing.

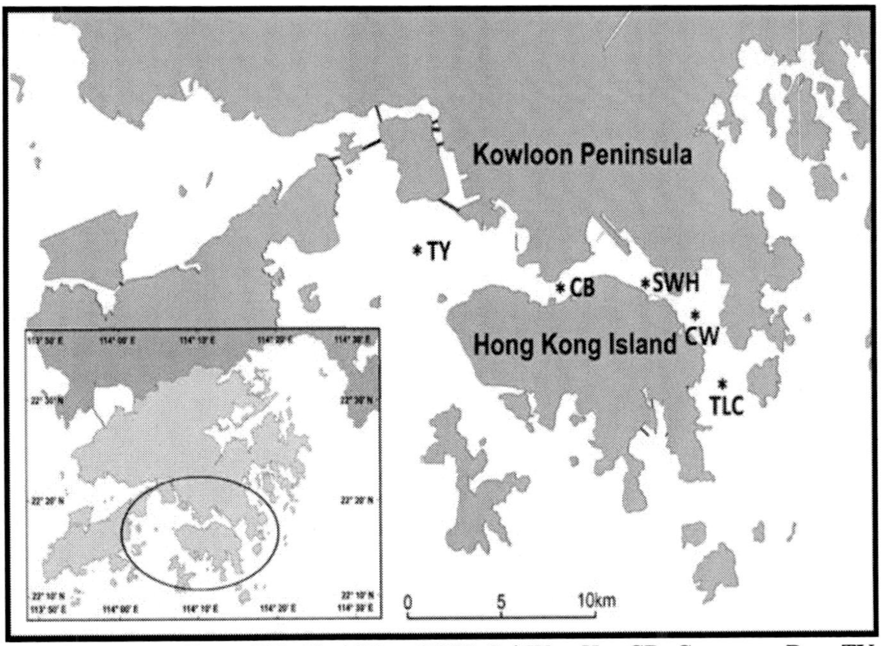

TLC: Tung Lung Chau; CW: Chai Wan; SWH: Sai Wan Ho; CB: Causeway Bay; TY: Tsing Yi.

Figure 1. Map of Victoria Harbour, Hong Kong showing the five sampling sites.

Laboratory Processing

Upon arrival to the laboratory, the sediment samples, except those for particle size analysis, were frozen at -20 °C for 24 hours followed by freeze-drying. The freeze-dried sediments were homogenized to <0.25 mm with a mechanical homogenizer and stored in a desiccator for further analyses. TOC in sediment was measured by Walkley-Black wet titration method (Walkley and Black 1934). TKN and TP in sediment were extracted by Kjeldahl acid digestion, followed by measurement using a Flow Injection Analyzer. Particle size distribution was determined with the wet sieving method and the data were expressed as median diameter (Md) in mm (Buchanan 1984).

Nematodes were extracted from the sediment by the Ludox™ centrifugation technique (Heip et al. 1985), using a 31 μm sieve for the lower size limit and a 500 μm sieve for the upper size limit. Each sample was washed into a lined Petri dish and the nematodes were counted and sorted under a stereoscopic microscope. Nematodes were slowly evaporated in anhydrous glycerol, mounted on permanent slides and identified to species or genus level using a compound microscope with bright-field illumination (Liu et al. 2007; Liu et al. 2011).

Nematode Biological Traits Analysis

Nematode species and genera were classified into five different biological traits according to their buccal morphology, tail shape, adult length, adult shape and life history strategy.

1) Buccal morphology. Based on the morphology of buccal cavity, Wieser (1953) classified the free-living nematodes into four feeding types: selective deposit feeder (1A), non-selective deposit feeder (1B), epigrowth feeder (2A) and predator (2B).
2) Tail shape. Tail shape was classified according to Thistle et al. (1995) into four types: short/round, elongated/filiform, conical and clavate.
3) Adult length and adult shape. Nematode species were assigned into four length groups: <1 mm, 1-2 mm, 2-4 mm and >4 mm. The nematode adult shape was classified according to Schratzberger et al. (2007) into three groups: stout (length-width ratio < 18), slender (length-width ratio 18-72), and long/thin (length-width ratio > 72).

4) Life history strategy. Nematodes were assigned to life history groups according to Bongers (1990) and Bongers et al. (1991, 1995) in which nematode genera were classified upon their ability for colonizing or persisting in a certain habitat. The five-point scale ranged from extreme colonizers (colonizer – persister, c-p score = 1) to extreme persisters (colonizer – persister, c-p score = 5).

Data Analysis

A) Sediment Parameters

For individual sediment parameters, data were analyzed for differences using two-way analysis of variance (ANOVA), with sites and seasons as fixed factors, followed by multiple comparison tests. For TOC, percent data were arcsine square-root transformed prior to analysis. All data were also subjected to normality and homogeneity of variance check using Kolmogorov–Smirnov and Levene's test, respectively.

B) Nematode Assemblages

For individual biological traits, percent data of different categories were arcsine square-root transformed and checked for normality and homogeneity of variance prior to analysis of differences using two-way ANOVA, with sites and seasons as fixed factors, followed by multiple comparison tests. For multivariate analysis of the nematode functional structure, a biological traits matrix was conducted by assigning each nematode taxon to each trait category according to its affinity. The biological trait was then raised by the relative species abundance to give an abundance-weighted traits matrix for each sample (Schratzberger et al. 2007). The abundance data were square-root transformed in order to down-weigh the contributions of quantitatively dominant species to the similarities calculated between samples (Clarke and Gorley 2006). Similarity between functional composition of nematode communities based on transformed abundance data was calculated based on Bray-Curtis coefficient (Bray and Curtis 1957) and analyzed using non-metric multidimensional scaling (MDS) (Clarke and Gorley 2006). Two-way analysis of similarities (ANOSIM) was performed to test differences among the five locations and the two sampling seasons. All the multivariate analyses were performed using the PRIMER 6 statistical package with the PERMANOVA+ add-on (PRIMER-E, Plymouth Marine Laboratory, UK).

RESULTS

Sediment Quality

Table 1 shows the range of sediment parameters from the two seasonal studies at the five sampling sites. The outer-harbour site TLC was the cleanest, in terms of nutrient levels, as compared to the other four sites. The central, inner-harbour site CB had a wider range of particle size spectrum, with a mix of fine and coarse sediments. It was also noted that the other outer-harbour site TY had nutrient levels intermediary between TLC and other inner-harbour sites.

Results of two-way ANOVA showed that there were significant spatial variations in the four sediment parameters among the sampling sites (Table 2). However, seasonal changes in sediment quality were minimal, except for significantly higher levels of TP noted in the wet, summer (August) than in dry, winter (January) season. Spatially, the inner-harbour site CB had coarser sediment and higher levels of TKN, TP and TOC than other sampling sites, whereas the outer-harbour site TLC contained the lowest nutrient content. The two inner-harbour sites SWH and CW and the other outer-harbour site TY had intermediary levels of TKN, TP and TOC between CB and TLC.

Nematode Biological Trait Analysis

A) Trophic Structure of Nematode Community

The feeding guild structure of the nematode communities sampled at the five locations in the two seasons is shown in Figure 2. The eastern outer-harbour site TLC had the lowest proportion of deposit feeders, 1A and 1B nematodes (27%), while the central harbour site CB had the highest proportion of 1A and 1B nematodes (80%). Also, TLC had the highest proportion of 2A nematodes, while CB had the lowest. All these sites had low proportion of 2B nematodes, while TLC had relatively higher proportion of 2B nematodes than that of the other sites.

Table 1. Range of sediment parameters for combined seasonal data (dried weight basis) at five sampling sites in Victoria Harbour

Parameter	TLC	CW	SWH	CB	TY
Median particle diameter (mm)	0.038-0.071	0.019-0.039	0.014-0.020	0.023-0.266	0.013-0.021
Total Kjeldahl nitrogen (mg/kg)	471.95-746.63	1101.88-1520.93	1364.89-2020.73	1888.24-2326.21	1026.52-1974.73
Total phosphorus (mg/kg)	337.82-467.52	571.02-712.70	635.50-847.72	789.71-883.08	630.51-774.03
Total organic carbon (% w/w)	0.26-0.56	0.95-1.17	1.56-1.89	1.43-2.00	1.03-1.51

Table 2. Results of two-way ANOVA and multiple comparisons for the sediment parameters at the sampling sites in January and August 2010

Sediment parameter	Factor	F value	*p* value	Multiple comparisons
Md	Site	11.084	0.000	(SWH<TY<CW<TLC)<CB
	Time	3.788	0.066	-
TKN	Site	50.723	0.000	TLC<(CW<[TY)<SWH]<CB
	Time	3.444	0.078	-
TP	Site	84.774	0.000	TLC<CW<(TY<[SWH)<CB]
	Time	8.619	0.008	Jan 2010<Aug 2010
TOC	Site	87.518	0.000	TLC<(CW<TY)<[SWH<CB]
	Time	1.627	0.217	-

Samples included in a same pair of parentheses, brackets or braces have no significant differences ($p<0.05$).

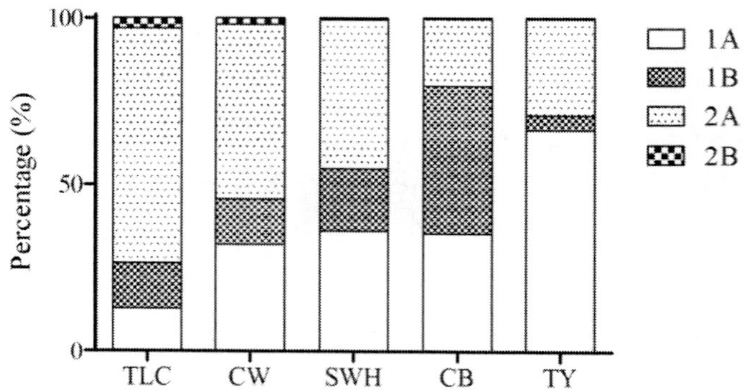

Figure 2. Feeding guild structure of nematodes at the five sampling sites in January and August 2010. 1A: selective deposit feeders; 1B: non-selective deposit feeders; 2A: epigrowth feeders; 2B: predators.

Results of two-way ANOVA showed that there were significant spatial differences for all the feeding type proportions, while there were no significant temporal differences (Table 3). Spatially, the eastern outer-harbour site TLC had the highest percentage of type 2A than that at the two inner-harbour sites CW and SWH, while the central harbour site CB and the western outer-harbour site TY had the lowest values. For deposit feeding nematodes (1A and 1B), TLC had significantly lower proportion than that at CB.

Table 3. Results of two-way ANOVA and multiple comparisons for the percentage of feeding types at five sampling sites in January and August 2010

Feeding type	Factor	F value	*p* value	Multiple comparisons
1A	Site	12.929	0.000	(TLC<CW<CB<SWH)<TY
	Time	2.343	0.141	-
1B	Site	13.163	0.000	(TY<CW<TLC<SWH)<CB
	Time	2.844	0.107	-
2A	Site	36.996	0.000	(CB<TY)<[SWH<CW]<TLC
	Time	2.081	0.165	-
2B	Site	7.218	0.001	(TY<CB<SWH<[CW)<TLC]
	Time	1.685	0.209	-
1A+1B	Site	16.839	0.000	(TLC<[CW<SWH)<{TY]<CB}
	Time	0.330	0.572	-

Samples included in a same pair of parentheses, brackets or braces have no significant differences ($p<0.05$).

B) Tail Shape

Over the two sampling seasons, the clavate tail shape was dominant at the five sampling sites. Their proportions were more than 40% at TLC, CW, SWH and CB. The elongated/filiform type (41.9%) was the second dominant tail shape, followed by conical (10.6%) and short/round type (0.5%). Most of the sampling sites had the proportions of elongated/filiform type more than 30%, whereas the short/round type had the lowest proportions at all the sampling sites (Figure 3).

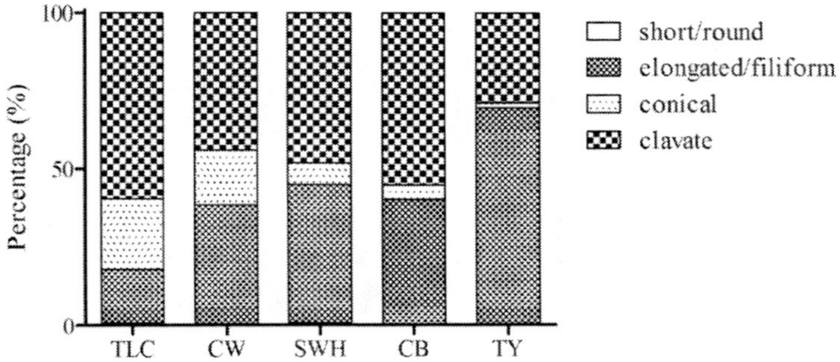

Figure 3. Tail shape structure of nematodes at the five sampling sites in January and August 2010.

Results of two-way ANOVA showed that there were significant spatial differences for most of the proportions of tail shape, and there was only significant temporal difference for the proportion of clavate type (Table 4). There was also no significant spatial and temporal difference for nematodes with short/round tail shape. For elongated/filiform type, the eastern outer-harbour site TLC had significantly lower proportion than that from CW, CB and SWH, while TY had the highest proportion. No significant temporal differences were found for this tail shape. For conical type, TLC had significantly higher proportion than that of TY and CB, but with no significant temporal differences. For clavate type, TLC and CB had significantly higher proportions than that of TY, while CW and SWH had intermediary proportions between the two groups. Temporally, samples in January 2010 had significantly higher proportions of clavate type than that in August 2010.

C) Adult Body Length

Over the two sampling seasons, most of nematodes sampled had the adult lengths in the range between 1-2 mm (48.5% at TLC, 65.4% at CW, 82.5% at SWH, 71.9% at CB and 94.5% at TY), followed by that with adult lengths in the range between 2-4 mm and that shorter than 1 mm. The proportions of nematodes with adult length longer than 4 mm were the lowest, and were only found at TLC (Figure 4).

Table 4. Results of two-way ANOVA and multiple comparisons for the percentage of tail shape at five sampling sites in January and August 2010

Tail shape	Factor	F value	*p* value	Multiple comparisons
Short/round	Site	0.573	0.685	-
	Time	4.243	0.053	-
Elongated/ filiform	Site	16.941	0.000	TLC<(CW<CB<SWH)<TY
	Time	4.031	0.053	-
Conical	Site	10.296	0.000	(TY<CB<[SWH)<{CW]<TLC}
	Time	2.646	0.119	-
Clavate	Site	6.085	0.002	(TY<[CW<SWH)<CB<TLC]
	Time	10.128	0.005	Aug 2010<Jan 2010

Samples included in a same pair of parentheses, brackets or braces have no significant differences ($p<0.05$).

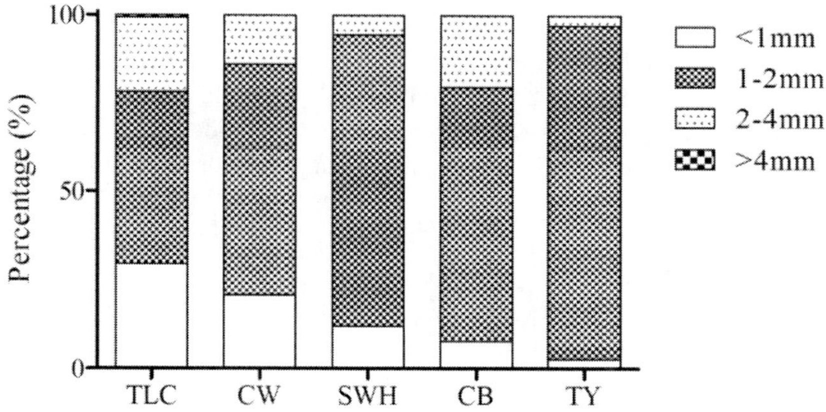

Figure 4. Adult body length structure of nematodes at the five sampling sites in January and August 2010.

Table 5. Results of two-way ANOVA and multiple comparisons for the percentage of adult body length at five sampling sites in January and August 2010

Adult length	Factor	F value	*p* value	Multiple comparisons
<1 mm	Site	9.556	0.000	(TY<[CB<SWH)<{CW]<TLC}
	Time	1.898	0.184	-
1-2 mm	Site	11.032	0.000	(TLC<[CW)<CB<{SWH]<TY}
	Time	0.909	0.352	-
2-4 mm	Site	4.138	0.013	(TY<[SWH<CW)<CB<TLC]
	Time	0.260	0.616	-
>4 mm	Site	74.014	0.000	(TY=CB=SWH=CW)<TLC
	Time	74.014	0.000	Jan 2010<Aug 2010

Samples included in a same pair of parentheses, brackets or braces have no significant differences ($p<0.05$).

Results of two-way ANOVA showed that there were significant spatial differences for all the proportions of the adult body length, and there was only significant temporal difference for the proportion of adult length more than 4 mm (Table 5). For the proportion of nematodes with adult length shorter than 1 mm, the eastern outer-harbour site TLC had significantly higher value than that of the inner-harbour sites SWH, CB and western outer-harbour site TY. For the proportion of nematodes with adult length between 1-2 mm, TLC had significantly lower value than that of CB, SWH and TY. For the proportion of nematodes with adult length between 2-4 mm, TLC and CB had significantly higher values than that of TY, and SWH and CW had the intermediary values between the two groups. For the proportion of nematodes with adult length more than 4 mm, only TLC had nematode samples with adult length in this range.

D) Adult Body Shape

Nematodes with slender shape were dominant in all the samples, with proportion more than 85% (Figure 5). The second dominant body type was stout shape, with proportion less than 4% in most of the samples, except that recorded in January 2010 at TLC. The proportions of the long/thin shape were the lowest, and were less than 1% in most of the samples, apart from those sampled in August 2010 at TLC and CW.

Results of two-way ANOVA showed that there were significant spatial differences for all the proportions of the adult body shapes, and only temporal differences for the proportions of stout and long/thin shape. The eastern outer-harbour site TLC had significantly higher proportions of stout and long/thin but lower proportion of slender nematodes than those of the other sites. Temporally, samples from August 2010 had significantly lower proportion of stout but higher proportion of long/thin nematodes than that from January 2010 (Table 6).

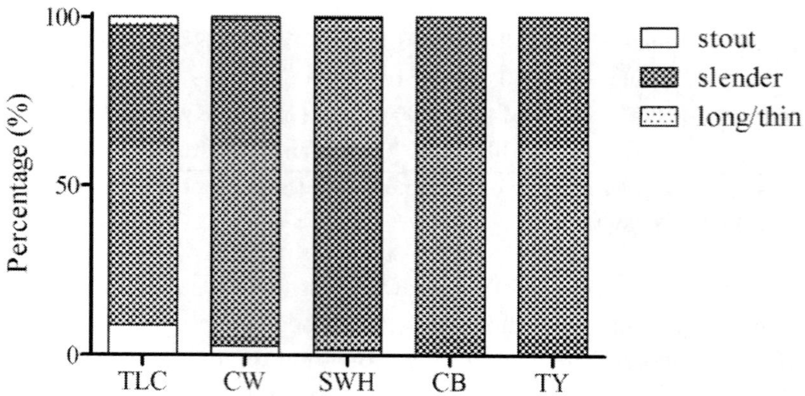

Figure 5. Adult body shape structure of nematodes at the five sampling sites in January and August 2010.

Table 6. Results of two-way ANOVA and multiple comparisons for the percentage of adult body shape at five sampling sites in January and August 2010

Adult body shape	Factor	F value	*p* value	Multiple comparisons
Stout	Site	24.425	0.000	(CB<TY<SWH<CW)<TLC
	Time	9.791	0.005	Aug 2010<Jan 2010
Slender	Site	15.681	0.000	TLC<(CW<[SWH<TY)<CB]
	Time	0.178	0.678	-
Long/thin	Site	11.031	0.000	(CB=TY<SWH<CW)<TLC
	Time	10.743	0.004	Jan 2010<Aug 2010

Samples included in a same pair of parentheses, brackets or braces have no significant differences ($p<0.05$).

E) Life History Strategy

The proportions of nematodes with c-p scale 1 and 5 were absent in all the samples. The proportions of nematodes with c-p scale 2 were the highest at most of the sampling sites, except for TY (31.7%). Their mean proportion was 79.9% at TLC, 62.5% at CW, 50.9% at SWH and 62.5% at CB. Site TY had the highest proportion of c-p 3 nematodes (67.8%), followed by SWH (47.1%), CB (36.7%), CW (33.9%) and TLC (12.0%). For proportion of nematodes with c-p scale 4, TLC had the highest value (8.1%), followed by CW (3.6%), SWH (2.1%), CB (0.8%) and TY (0.4%) (Figure 6).

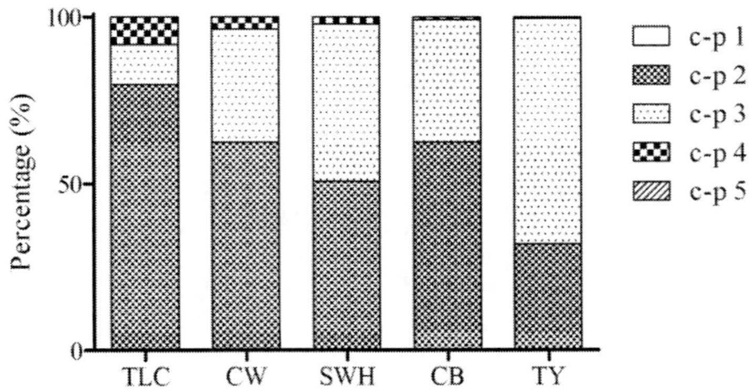

Figure 6. Life history strategy structure of nematodes at the five sampling sites in January and August 2010.

Results of two-way ANOVA showed that there were significant spatial differences for the existed three c-p scales, but without temporal differences. For proportion of nematodes with c-p scale 2, TLC had significantly higher value than that of TY, while SWH, CW and CB had intermediary values between the two groups. For proportion of nematodes with c-p scale 3, TY had significantly higher value than that from the other sites, whereas for proportion of nematodes with c-p scale 4, TLC had significantly higher value than those of the other sites (Table 7).

F) Biological Traits Analysis of Nematode Community

The MDS ordination based on biological trait combination separated the nematode communities from the eastern outer-harboue site TLC from those inner-harbour sites (CW, SWH, CB), whereas the western outer-harbour TY fell in between (Figure 7). Results of two-way ANOSIM test based on the sampling sites and time showed that there were significant spatial (Global R=0.485, p=0.001) and temporal differences (Global R=0.4, p=0.001). Significant differences were also found between most of the sampling sites, except between CW and SWH (R=0.037, p=0.42) (Table 8).

Table 7. Results of two-way ANOVA and multiple comparisons for the percentage of life history strategy (c-p scale) at five sampling sites in January and August 2010

c-p scale	Factor	F value	*p* value	Multiple comparisons
c-p 2	Site	4.609	0.008	(TY<[SWH<CW<CB)<TLC]
	Time	0.848	0.368	-
c-p 3	Site	15.708	0.000	(TLC<[CW)<CB<SWH]<TY
	Time	1.149	0.297	-
c-p 4	Site	10.029	0.000	(TY<CB<SWH<CW)<TLC
	Time	1.012	0.326	-

Samples included in a same pair of parentheses, brackets or braces have no significant differences ($p<0.05$).

DISCUSSION

Sediment Quality

In the present study, levels of TOC, TKN and TP in the inner-harbour area were still significantly higher than that in the outer-harbour area, which suggested that organic enrichment of the sediments caused by previous sewage discharges persisted even after the abatement of such pollution since 2001. The low recovery rate in sediment nutrients could be due to the mixture of discharges from urban and industrial sources, which may not be readily bio-available to microorganisms for degradation in the harbour (Liu et al. 2011). Heavy metals tend to be adsorbed on the surface of sediment particles, with higher concentrations of metals associating with smaller particles (Tang et al. 2008). The presence of these toxic heavy metals may also inhibit the degradation process by microorganisms, thus reducing the recovery of the sediment quality in the habour area (Fabiano et al. 1994).

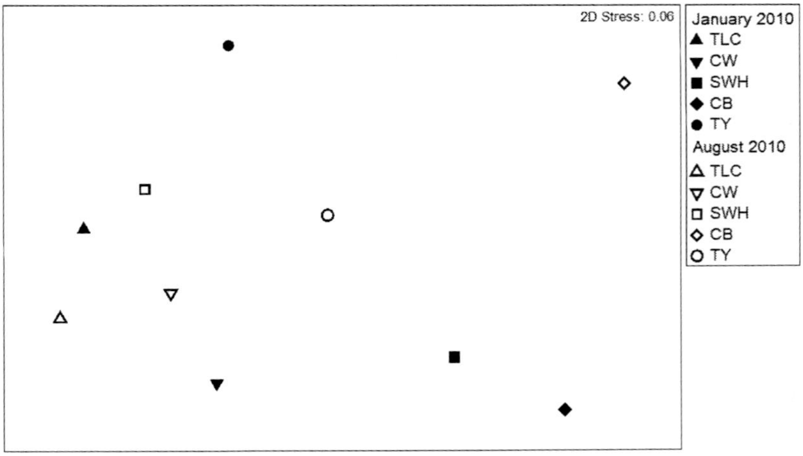

Figure 7. Non-metric multi-dimensional scaling (MDS) ordination based on mean biological trait combinations of nematode communities at the five sampling sites and two seasons.

Table 8. Results of two-way ANOSIM for biological trait combinations of nematode communities based on sampling sites and time

Groups	R Statistic
TLC, CW	0.315*
TLC, SWH	0.593*
TLC, CB	0.944*
TLC, TY	0.833*
CW, SWH	0.037
CW, CB	0.556*
CW, TY	0.315*
SWH, CB	0.444*
SWH, TY	0.463*
CB, TY	0.556*
Global R	0.485*
January 2010, August 2010	0.400*

*Significant level < 5%.

Nematode Assemblages

Organic materials in the sediment can serve as a major source of nutrients for the meiofauna, including nematodes (Coull and Chandler 1992). In the present study, the proportion of the deposit feeders (1A and 1B) at each sampling site was highly correlated with the nutrient level in the harbour sediments. The central harbour site CB and the western harbour site TY had higher proportions of deposit feeders, while the eastern outer-harbour site TLC had relatively higher proportion of epigrowth feeders. This may be due to the different food sources available in the inner- and outer-harbour areas. Victoria Harbour has been suffered from sewage discharge for the past decades, and the sediments, especially those in the inner-harbour area, are still organically enriched (Liu et al. 2011). The elevated levels of nutrients can be a food source for nematodes, resulting in the dominance of deposit feeding nematodes in these areas. Pallo et al. (1998) reported that deposit feeding nematodes were found dominant in the Gulf of Riga, where is considered to be one of the most eutrophic areas in the Baltic Sea. Unlike the case for the proportion of the deposit feeders, the proportion of the predatory nematodes at the eastern outer-harbour site TLC was significantly higher than the other sites. This situation was also reported by Pallo et al. (1998) that predatory nematodes were only found in higher abundance in sediments with low organic content. Considering the proportions of deposit feeders and predators in Victoria Harbour, it can be concluded that the inner, central harbour area was still more polluted than the outer-harbour area.

Buccal morphology has been widely used in ecological studies to classify the nematodes by feeding types, which is a useful functional attribute for nematode communities (Schratzberger et al. 2007). Tail shape morphology is another important functional classification, as they are regarded to play a crucial role in locomotion, foraging and reproduction (Platt et al. 1983; Thistle et al. 1985; 1995). Nematodes with short tails are thought to have a mobile lifestyle, while nematodes with long tails are thought to be less mobile or even hemisessile (Riemann 1974). Nematodes with clavate tail shape were dominant at most of the sampling sites, and there were no significant spatial differences among sites TLC, CW, SWH and CB. This indicated that nematodes with clavate tail shape were widely distributed in the study area, and can live in both polluted and unpolluted areas. The proportions of nematodes with conical tail shape were higher at eastern outer-harbour site TLC than those at sites CB and TY, which suggested that this kind of nematodes may likely to live in less polluted areas. The proportions of nematodes with elongate/filiform tail shape were found higher in more polluted areas, which implied that this kind of nematodes was likely to live in polluted areas. One possible implication is that the elongated tail of nematodes can enlarge the surface area of the body which in turn can increase the oxygen absorption efficiency per unit of volume when living in the sediment (Soetaert et al. 2002). However, nematode tail shapes are mainly related to their adaptation to sediment texture (Riemann 1974), so it is not always accurate to use tail shapes of nematodes to reflect *in-situ* pollution conditions.

In addition to tail shape, the adult length and body shape of nematodes can provide functional information of the nematode communities (Soetaert et al. 2002). The body shape is assumed to be an adaptation to a certain life style, with the long/thin body shape moving swiftly through the sediment to escape from predation than those with stout body shape (Tita et al. 1999; Soetaert et al. 2002). The nematode body size distribution is also considered to be determined by food availability, with a pattern that reduction in food availability resulting in a decrease in nematode body size. In the present study, the proportion of nematodes with body length lower than 1 mm was significantly higher at the eastern outer-harbour site TLC than those at sites TY, CB and SWH, which may imply that in the inner-harbour area, the higher organic content in sediments provides more abundance of food. For the body shape, the proportions of stout and long/thin nematodes were significantly higher at TLC than those at the other sites. Slender nematodes were dominant at all the sites, which indicated that this kind of nematodes was the most adaptive group in the benthic environment of Victoria Harbour and its vicinity. The eastern outer-harbour site TLC had more diverse body shapes, reflecting the higher diversity of nematode species within the benthic meiofaunal communities. Under disturbances such as sewage discharge, the community equilibrium is shifted since the rate of community damage is faster than the rate of community recovery (Warwick 1986). In such case, the *K*-selected species are less tolerant and die, while the *r*-selected species become dominant in abundance and biomass. In the present study, marine nematode taxa were classified on a five-point colonizer-persister (c-p) scale 1−5, according to Bonger (1990) and Bonger et al. (1991; 1995). The colonizers (c-p scale ≤ 2) can be referred to as *r*-selected species, which are characterized by short life-cycle, and high reproduction rate, colonization ability and tolerance to various types of disturbances. The higher proportion of colonizers may also indicate that the environment is under continuous perturbations (Bonger et al. 1991). In contrast, the persisters (c-p scale >3) can be referred to as *K*-selected species, which are characterized by long life-cycle, low colonization ability and few offspring, and are sensitive to disturbances. In marine sediments, the extreme colonizers (c-p scale of 1) and persisters (c-p scale of 5) are virtually absent (Schratzberger et al. 2007). Although there were significant spatial differences for proportions of nematodes with different c-p scales in Victoria Harbour, no clear spatial patterns which can separate the polluted inner-harbour from cleaner outer-harbour sites could be detected. For example, no spatial difference was found for the proportion of nematodes with c-p scale of 2 (*r*-selected species) among the cleaner eastern outer-harbour site TLC and the inner-harbour sites CW, SWH and CB. For proportion of nematodes with c-p scale of 3 (*K*-selected species), TLC even had significantly lower value than those at SWH and CB, where higher organic levels were found. Hence, in the present study, the structure of nematode with different c-p scales at each sampling site cannot be used to reflect the degree of organic enrichment in the harbour areas. Further investigations are required to qualify this classification as an indicator of pollution.

CONCLUSION

Schratzberger et al. (2007) firstly proposed to use a range of biological traits of nematode species to identify patterns in the functional composition of their assemblages in temperate waters. The present study demonstrated that such an approach can also be applicable in analysis of nematode functional groups based on biological traits in subtropical harbour environment. Results showed that the benthic environment at the inner-harbour sampling sites was still polluted even after more than 10 years of cessation in sewage discharge, according to data from individual biological trait attributes and a combined analysis. However, data from some attributes, such as the colonizer-persister scale, were not consistent with the above observations. The present findings thus suggested that biological trait analysis can be complementary to community composition analysis in deciphering the response of nematode communities from the cessation of sewage pollution, while the use of a single trait or index should be cautioned.

ACKNOWLEDGMENTS

Thanks for Ankang Teng for his technical assistance. This study is fully supported by a General Research Fund (Project No. CityU 161009) from the Research Grants Council of the HKSAR Government.

REFERENCES

Bongers, T., 1990. The maturity index: an ecological measure of environmental disturbance based on nematode species composition. *Oecologia* 83, 14-19.

Bongers, T., Alkemade, R., Yeates, G.W., 1991. Interpretation of disturbance-induced maturity decrease in marine nematode assemblages by means of the Maturity Index. *Marine Ecology Progress Series* 76, 135-142.

Bongers, T., de Goede, R.G.M., Korthals, G.W., Yeates, G.W., 1995. Proposed changes of c-p classification for nematodes. *Russian Journal of Nematology* 3, 61-62.

Boyd, S.E., Rees, H.L., Richardson, C.A., 2000. Nematodes as sensitive indicators of change at dreged material disposal sites. *Estuarine, Coastal and Shelf Science* 51, 805-819.

Bray, J.R., Curtis, J.T., 1957. An ordination of an upland forest community of Southern Wisconsin. *Ecological Monograph* 27, 325-349.

Bremner, J., Rogers, S.I., Frid, C.L.J., 2006. Methods for describing ecological functioning of marine benthic assemblages using biological traits analysis (BTA). *Ecological Indicators* 6, 609-622.

Buchanan, J.B., 1984. Sediment analysis. In, Holme, N.A., McIntyre, A.D. (Eds.), Methods for the Study of Marine Benthos. IBP Handbook (second ed.), vol. 16. Blackwell Scientific Publications, Oxford, UK, pp. 41-65.

Clarke, K.R., Gorley, R.N., 2006. PRIMER v6: User Manual/Tutorial. PRIMER-E, Plymouth, UK, pp. 190.

Coull, B.C., 1999. Role of meiofauna in estuarine soft-bottom habitats. *Australian Journal of Ecology* 24, 327-343.

Coull, B.C., Chandler, G.T., 1992. Pollution and meiofauna: field, laboratory, and mesocosm studies. *Oceanography and Marine Biology: An Annual Review* 30, 191-271.

HKEPD, 2004. *Annual Marine Water Quality Reports*. Environmental Protection Department, Hong Kong SAR Government.

Essink, K., Keidel, H., 1998. Changes in estuarine nematodes communities following a decrease of organic pollution. *Aquatic Ecology* 32, 195-202.

Fabiano, M., Danovaro, R., Magi, E., Mazzucotelli, A., 1994. Effects of heavy metals on benthic bacteria in coastal marine sediments: A field result. *Marine Pollution Bulletin* 28, 18-23.

Fraschetti, S., Gambi, C., Giangrande, A., Musco, L., Terlizzi, A., Danovaro, R., 2006. Structural and functional response of meiofauna rocky assemblages to sewage pollution. *Marine Pollution Bulletin* 52, 540-548.

Gee, J.M., 1989. An ecological and economical review of meiofauna as food for fish. *Zoological Journal of the Linnaean Society* 96, 243-261.

Giere, O., 1993. Meiobenthology - the microscopic fauna in aquatic sediments. Springer - Verlag, Berlin, Germany, pp. 328.

Guo, Y.Q., Warwick, R.M., Zhang, Z.N., Mu, F.H., 2002. Free-living marine nematodes as a pollution indicator of the Bohai Sea. *Journal of Environmental Sciences - China* 14, 558-562.

Gyedu-Ababio, T.K., Furstenberg, J.P., Baird, D., Vanreusel, A., 1999. Nematodes as indicators of pollution: a case study from the Swartkops River system, South Africa. *Hydrobiologia* 397, 155-169.

Heip, C., 1980. Meiobenthos as tool in the assessment of marine environmental quality. Rapport et Procès-verbeaux des Réunions. *Conseil International de l'Exploration de la Mer, Copenhague* 179, 182–187.

Heip, C., Vincx, M., Vranken, G., 1985. The ecology of marine nematodes. *Oceanography and Marine Biology: An Annual Review* 23, 399-499.

HKEPD (2004). Annual Marine Water Quality Reports. Environmental Protection Department, Hong Kong SAR Government.

Jensen, P., 1987. Differences in microhabitat, abundance, biomass and body size between oxybiotic and thiobiotic free-living marine nematodes. *Oecologia* 71, 564-567.

Liu, X.S., Xu, W.Z., Cheung, S.G., Shin, P.K.S., 2008. Subtropical meiobenthic nematode communities in Victoria Harbour, Hong Kong. *Marine Pollution Bulletin* 56, 1491-1497.

Liu, X.S., Xu, W.Z., Cheung, S.G., Shin, P.K.S., 2011. Marine meiobenthic and nematode community structure in Victoria Harbour, Hong Kong upon recovery from sewage pollution. *Marine Pollution Bulletin* 63, 318-325.

Liu, X.S., Zhang, Z.N., Huang, Y., 2007. Sublittoral meiofauna with particular reference to nematodes in the southern Yellow Sea, China. Estuarine, Coastal and Shelf Science 71, 616-628.

Morton, B., Morton, J., 1983. The Sea Shore Ecology of Hong Kong. Hong Kong University Press, Hong Kong, pp. 350.

Pallo, P., Widbom, B., Olaffson, E., 1998. A quantitative survey of the benthic meiofauna in the Gulf of Riga (Eastern Baltic Sea) with special reference to the structure of nematode assemblages. *Ophelia* 49, 117-139.

Peters, R.H., 1983. The ecological implications of body size. Cambridge University Press, Cambridge, UK, pp. 329.

Platt, H.M., Shaw, K.M., Lambshead, P.J.D., 1984. Nematode species abundance patterns and their use in the detection of environmental perturbations. *Hydrobiologia* 118, 59-66.

Platt, H.M., Warwick, R.M., 1983. Free-living marine nematodes. Part I. British Enoplids. In: Synopses of the British Fauna (New Series), No. 28. Cambridge University Press, Cambridge, UK, pp. 307.

Riemann, F., 1974. On hemisessile nematodes with flagelliform tail living in marine soft bottoms and on micro-tubes found in deep sea sediments. *Mikrofauna Meeresbodens* 40, 1-15.

Schratzberger, M., Warr, K., Rogers, S.I., 2007. Functional diversity of nematode communities in the southwestern North Sea. *Marine Environmental Research* 63, 368-389.

Soetaert, K., Muthumbi, A., Heip, C. 2002. Size and shape of ocean margin nematodes: morphological diversity and depth-related patterns. *Marine Ecology Progress Series* 242, 179-193.

Statzner, B., Resh, V.H., Roux, L.A., 1994. The synthesis of long-term ecological research in the context of concurrently developed ecological theory: design of a research strategy for the Upper Rhone River and its floodplain. *Freshwater Biology* 31, 253-263.

Tang, C.W.Y., Ip, C.C.M., Zhang, G., Shin, P.K.S., Qian, P.Y., Li, X.D., 2008. The spatial and temporal distribution of heavy metals in sediments of Victoria Harbour, Hong Kong. *Marine Pollution Bulletin* 57, 816-825.

Thistle, D., Lambshead, P.J.D., Sherman, K.M., 1995. Nematode tail-shape groups respond to environmental differences in the deep sea. *Vie et Milieu* 45, 107-115.

Thistle, D., Sherman, K.M., 1985. The nematode fauna of a deep-sea site exposed to strong near-bottom currents. *Deep-Sea Research* 32, 1077-1088.

Tita, G., Vincx, M., Desrosiers, G., 1999. Size spectra, body width and morphotypes of intertidal nematodes: an ecological interpretation. *Journal of the Marine Biological Association of the United Kingdom* 79, 1007-1015.

Vanaverbeke, J., Soetaert, K., Vincx, M., 2004. Changes in morphometric characteristics of nematode communities during a spring phytoplanton bloom deposition. *Marine Ecology Progress Series* 273, 139-146.

Vanaverbeke, J., Steyaert, M., Vanreusel, A., Vincx, M., 2003. Nematode biomass spectra as descriptors of functional changes to human and natural impact. *Marine Ecology Progress Series* 249, 157-170.

Walkley, A., Black, I.A., 1934. An examination of the Degtjareff method for determining soil organic matter, and a proposed modification of the chromic acid titration method. *Soil Science* 37, 29-38.

Warwick, R.M., 1986. A new method for detecting pollution effects on marine macrobenthic communities. *Marine Biology* 92, 557-562.

Wieser, W., 1953. Die Beziehung zwischen Mundhöhlengestalt, Ernährungsweise und Vorkommen bei freilebenden marinen Nematoden. *Arkiv för Zoologi* 4, 439-484.

In: Nematodes
Editor: Lee M. Davis

ISBN: 978-1-62948-764-9

Chapter 2

LECTINS AND NEMATODES: AN ALTERNATIVE CONTROL STRATEGY IN AGRICULTURE

José Thalles J.G de Lacerda*, Tatiane Santi Gadelha and Carlos Alberto A. Gadelha

Department of Molecular Biology, Laboratory of Biochemistry, Genetics and Radiobiology (BIOGER), Federal University of Paraíba. João Pessoa, Paraíba, Brasil. Cidade Universitária, João Pessoa, Paraíba – CEP, Brazil

ABSTRACT

The Nematoda phylum is a numerous and diversified group of metazoans. According to estimates, it includes between 100,000 and 1,000,000 free-living (terrestrial or aquatic) and parasitic species that are found parasitizing animals (vertebrates and invertebrates) as well as plants, with beneficial or harmful effects. Entomoparasitic nematodes (EPNs) living symbiotically with bacteria are used in biological control strategies against insects in agriculture. Sedentary phytonematodes like Meloidogyne incognita are among some of the main agricultural pathogens, responsible for annual losses accounting to US$125 billion a year. Currently, control strategies against these parasites rely on the use of chemical vermicides that may not be as efficient as they once were.

---

* Corresponding author: E-mail: thalles_lacerda2@hotmail.com.

Apart from this, these chemicals are harmful to the environment, extending their effects to other species, even mammals. In this sense, new and more viable control strategies that are less environmentally noxious have become increasingly important. In this sense, more recently lectins have been successfully used as a new model in the study of nematodes. Lectins are non-immune hemagglutinating proteins or glycoproteins constitutively expressed in seeds, leaves, barks, roots, tubers and fruit of several plant species, though these compounds may also be found in all kinds of organisms. They are biodegradable, and therefore do not build up in the soil or in organisms that feed on species controlled with lectins. This chapter describes how lectins may Interact with nematodes, reviews assays of the use of plant lectins to control phytonematodes, and how lectins are used by EPN.

Keywords: Nematodes; lectins; agriculture

Introduction

The Nematoda phylum is a large, diversified group of metazoans. Nematoda species exhibit morphological characteristics in common, such as an elongated and cylindrical body, as well as the presence of pseudocoelom (Wood, 1988). Some nematode species exhibit free-living habits, both terrestrial and aquatic, while other species are parasititic. Parasite nematodes are detected in both animal (vertebrae and invertebrate) animals and plants. Total nematode species are estimated between 100,000 and 1,000,000 (Lambshead, 1993). The main sensory organs of nematodes are the lip papillae and amphyds, located in the lip area (Ferraz & Monteiro, 1995; Perry & Aumann, 1998; Wright, 1981). Apart from sensory organs, carbohydrates on the cuticle surface of nematodes play an essential role in the specific interaction between the parasite and its host (Zuckermann & Jonsson, 1984). Higher carbohydrate concentrations on the cuticle indicate an increased perception of the molecular signal emitted by a given chemical compound by the receptor (Zuckermann, 1983).

Nematodes, insects, fungi, viruses and bacteria are responsible for devastating damage in agriculture, causing considerable economic losses. In this scenario, sedentary endoparasitic phytonematodes are the main phytopathogens affecting agriculture worldwide, and lead to losses estimated at US$125 billion a year (Chitwood, 2003). For Williamson & Gleason (2003), the main phytoparasitic nematodes are those that cause cysts (*Heterodera* spp.and *Globodera* spp.) and species that form root knot galls (*Meloidogyne* spp.). *Meloidogyne* is considered the most harmful genus in agriculture activities around the world, accounting for 80% of losses (Williamson e Kumar, 2006), and *M. incognita* is known to infect around 2,000 plant species (Trudgill and Blok, 2001) (Figure 1).

In Brazil, approximately 100 nematode species have been recorded in association with 600 plant species (Embrapa, 2007). In nematodes, parasitism is ruled by proteins and other molecules secreted that, combined with the mechanical action exerted by the stylet, promote the penetration and migration of the nematode to the host plant's roots. These effectors play a role in the prevention of infection and in the regulation of the plant's defense mechanism, apart from starting and keeping its feeding site. Sedentary endoparasitic phytonematodes establish a very close and sophisticated relationship with host plants (Abad et al., 2003), promoting the invasion of the plant tissue, evasion of defense mechanisms, and formation of permanent feeding sites (Davis et al., 2004) (Figure 2). The kind and distribution of carbohydrate residues on the body of nematodes vary across species, indicating the specific character towards chemical attraction signals (Zuckerman, 1983). Two types of receptors have been described: chemoreceptors associated with the neuron membrane, and surface receptors – both of which have carbohydrates in their structures (Jonsson, 1987). After the external chemical stimulus reaches the sensory organs, molecules interact with neighboring cells, transmitting the signal through nervous cells to the central nervous systems, where the response is processed and converted in a behavioral action (Jonsson, 1987).

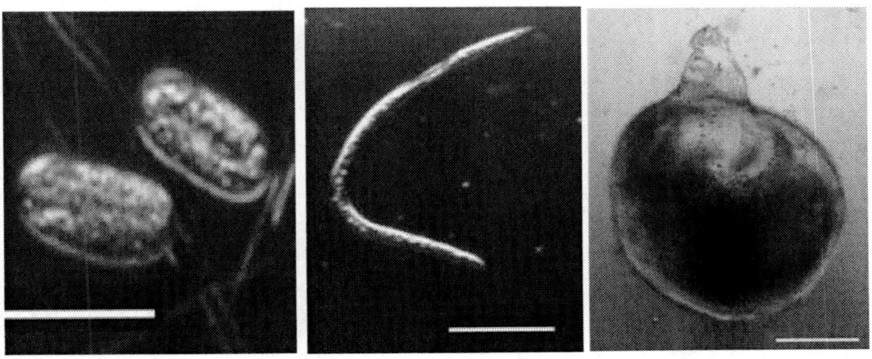

Figure 1. Secondary endoparasitic nematodes. a) *Meloidogyne incognita* eggs. b) J2 stage, infectious form of the species. c) Adult *M. incognita* adult female (Photos kindly provided by Dr. José Dijair Antonino Jr.).

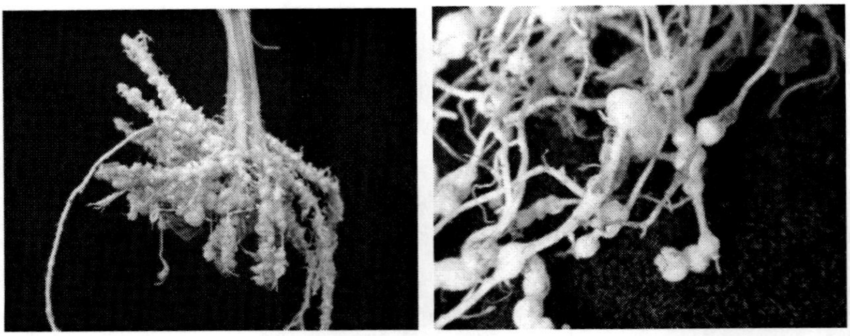

Figure 2. Destruction of roots caused by secondary endoparasitic phytonematodes. a) Cucumber roots infected by *M. incognita*. b) Galls formed on lettuce roots by *M. incognita* (Photos kindly provided by Dr. Carlos Eduardo Rossi).

The structure and composition of nematode communities are influenced by changes in the vegetation they colonize (Pattison et al., 2008), in soil (Gupta & Yeates 1997; Yeates & Pattison, 2006), and in environmental conditions, either natural or anthropogenic. More specifically, the physical properties of soils, such as texture, water content and structure may affect nematode assemblages directly or indirectly (Yeates & Bongers 1999; Kandji et al., 2001; Gomes et al., 2003; Kimenju et al., 2009).

Currently, control strategies against nematodes comprise mainly the use of synthetic chemical nematicides. Being highly toxic, these compounds are critically harmful to the environment, contaminating soils, water bodies and other species as well, like mammals (Freitas et al., 2001). Nematicidal efficiency is quite low, since parasitic phytonematodes exhibit endogenous character, which protects these animals on plant roots for most part of their lifecycle (Trudgill & Blok, 2001).

In this sense, secondary metabolites have been used to fight phytonematodes. These molecules are known as natural insecticides. They may confuse insects or even kill them, and present the interesting advantage of being consistently specific (Vieira et al., 2001). Seed extracts may also exert nematicidal as well as nematosatic effects, in which case these compounds temporarily inhibit nematode development in a specific way (Wall et al., 1999). Such effects are associated with lectins and other secondary metabolites. As biodegradable molecules that are unable to build up in soils or in organisms upstream the food chain that feed on organisms that are fought using lectin-based control strategies, lectins are increasingly used in agriculture.

Lectins are non-immune proteins or glycoproteins that reversibly bind to specific carbohydrates and play an array of biological activities. These substances have been associated with the blockade of chemoreceptors of the nematode, making it difficult for the organism to move and locate a potential host plant (Marban-Mendonza et al., 1987). The binding of an attractive chemical released by plant roots to the terminal carbohydrate recognition site of the glycoprotein receptor causes a normal chemotactic interaction. This phenomenon is blocked by lectins, which bind to the terminal carbohydrate recognition site; alternatively, lectins may completely block this chemotactic interaction based on the action of hydrolytic enzymes (Jonsson, 1987).

However, some nematodes are actually useful. Nematode species have been used as adjuvant in biological control strategies to colonize arthropods. These species are called entomopathogenic nematodes, and are defined as organisms with the potential to kill or cause diseases in insects. The terms entomopathogenic is widely used in biology to refer to organisms that release toxins and other substances with insecticide properties (Ortiz-Urquiza et al., 2010; Campos et al., 2010; Toledo et al., 2010). Nematode species associated with insects distributed across a variety of families, four of which are consistently important and interesting in control strategies, from the economic perspective: Mermithidae, Sphaerulariidae, Steinernematidae and Heterorhabditidae (Popiel & Hominick, 1990).

The use of nematodes in agriculture stands as a viable alternative due to their potential to disperse in the soil and actively search for a host, as well as to the ability of recycling using the host itself, and easy laboratory management (Bathon, 1996; Ehlers & Hokannen, 1996; West & Vrain, 1997; Medeiros et al., 2000).

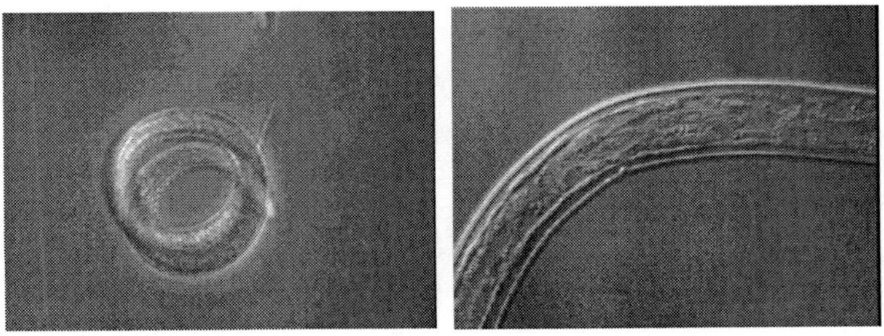

Figure 3. a) curled JI (*Heterorhabditis baujardi* LPP7). Presence of Double cuticle. b) Rod-shaped bacilli (*Photorhabdus*) next to one another in the digestive tube of the JI *Heterorhabditis baujardi* in a symbiosis process (Photos kindly provided by Dr. Carlos Eduardo Winter).

The symbiosis between some of these nematodes and bacteria may be quite specific: the genera *Heterorhabditis* and *Steinernema* are exclusively colonized by bacteria of the genera *Photorhabdus* and *Xenorhabdus*, respectively (Poinar, 1990) (Figure 3). The efficiency with which the nematode-bacterium complex kills a host varies across species, physiological and immunological status of the insect affected, and on particular aspects of the bacterium species used (Akhurst & Dunphy, 1993).

In the lifecycle of an entomoparasitic nematode (EPN), bacterial symbionts play an essential role: to kill the insects and allow the nematode to feed on the dead organism, apart from controlling the established microecosystem and blocking the infection by other microorganisms. Entomopathogenic bacteria have an extensive repertoire of genes involved in the production of toxins and secondary metabolites. It is believed that these molecules guarantee the pathogenicity process against the insect and the recognition by the nematode and maintenance of symbiosis (Clarke, 2008). Lectins secreted by symbionts like the bacterium *X. nematophila* and the nematode *S. carpocapsae* have been associated with processes like insect invasion and survival to immune response, mainly due to the binding to the mucosa, based on functional analyses conducted by Heugens et al. (2002) that allowed the identification of mutated genes.

Some plant lectins have been proved to exert entomotoxic action when ingested by insects of the Coleoptera, Homoptera and Lepidoptera orders. These lectins bind to the enzymes present on the brush-border of an insect's midgut epithelial cells, to the quitin in the peritrophic or perimicrovilar membrane, or to glycolized digestive enzymes (Chrispeels & Raikhel, 1991; Peumans & Van Damme, 1995).

THE GLYCOSYLATION OF NEMATODES

Nematodes synthesize uncommon glycoproteins and carbohydrate binding proteins (GBPs) that may play an antigenic role or be involved in the invasion, lodging and parasitism. Their survival strategies usually admit the participation of glycoconjugates that form a protective barrier against the host's immune system. To understand the glycosylation of nematodes, we will discuss *Caenorhabditis elegans*, a species believed to be an ideal candidate for being easily managed genetically and in culture, and to the fact that they were the first to be used in the study of apoptosis and the use of interference RNA in gene silencing. This species' genome codifies several enzymes implicated in the biosynthesis of glyconjugates, such as the enzyme responsible for the synthesis of S-GalNAc, N-glycans, glycosaminoglycans, glycosphingolipids e glycosylphosphatidylinositol, though their carbohydrate profile changes with cell cycle stages (Cummings & Esko, 2009).

C. elegans and other nematodes might be expected to hold several aspects in common in terms of structures and glycoconjugate synthesis, though individual nematode species differ regarding their carbohydrates (Figure 4).

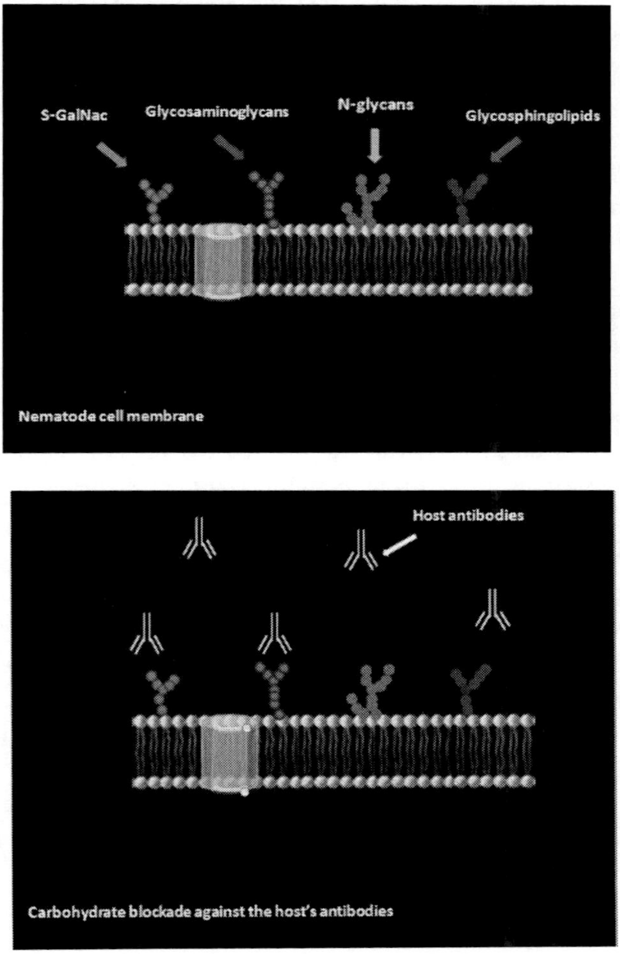

Figure 4. Glycoconjugates of a nematode. a) Description of some glycoconjugates in the cell membrane of *C. elegans*. b) Use of the carbohydrate layer to block the action of host's antibodies and facilitate lodging and parasitism.

Carbohydrates like $Man_5GlcNAc_2$-Asn are cleaved by α-glycosidades I and II and α-manosidades I and II. It is modified by *N*-acetylglycosamine, generating $GlcNAc_1Man_5GlcNAc_2$-Asn. Therefore, β-hexosaminidase, α-manosidades, α-fucosyltransferases form N-glycans with three or four mannose residues and a nucleus of α1-3 fucose residues. Nematodes may produce more complex structures, such as phosphorylcolin bound to the residues of chitinose *N*-acetylglycosamine in the center. This is modification that seems to be a common characteristic of nematode glycoproteins. Their O-glycans present uncommon structures such as the binding of GlcAβ-3GalNAc associated with O-glycans containing β-glycose and residues of glycoronic acid, residues of α1-2 fucose, and O-methylated 2 fucose. Its glycosphyngolipids are composed by GlcNAcβ1-3Manβ1-4Glcβ1 (Cummings & Esko, 2009).

C. elegans synthesize several lectins, among which galactin, the first glycoprotein identified in these organisms. They produce 135 type C lectins (CLTD), though only 19 have carbohydrate recognition sequences, and most do not present transmembrane domains. This indicates that they act as signaling proteins. Studies have shown that their glycolipids may be responsible for the resistance to toxins produced by bacteria that would lead to its death, such as the toxin *Bacillus thuringiensis* produces. Mutagenic nematode species do not seem to be affected by toxins (Cummings& Esko, 2009). Glycolipids also play a role in facilitating the EPN:bacterium symbiosis process.

The presence of these glycoconjugates in the membrane of nematodes promotes recognition by leptons. For this reason these molecules, by recognizing specific carbohydrates, may be used as nematicides or as biological markers for certain nematodes. Different nematode species present distinctive membrane carbohydrate profiles, a distinction that imparts specific characteristics regarding their hosts, their parasitism mode, and the way these nematodes may be infected and eliminated (Table 1).

Table 1. Some glycoconjugates that are or not synthesized by *C. elegans*

Glycoconjugates	Synthesized	Non-synthesized
Heteropolysaccharide	Heparin-sulphate	Keratan-sulphate
	Chondroitin (no sulphate)	Dermatan-sulphate
Lectins	Galectins Type C lectins Type L lectins Type R lectins	Type I lectins

The structural flexibility of carbohydrates testifies for the glycolisation profile as an individual characteristic. In this sense, sugars may effectively be considered as the main mechanism that stores information.

The Plant-Nematode Interaction

Sedentary phytonematodes establish long-term relationships with their hosts. Root vascular cells are converted in large multinucleate cells responsible for nutrition, which nematodes use to feed for about one month. The most studied – and complex – nutrition structures are the synsytia and giant cells induced by cyst and gall-producing nematodes, respectively (1981). The cell cycle and the fragmentation of cytoskeleton filaments are activated, which are crucial in the development of nutrition cells (De Almeida-Engler et al., 1999). The generation of giant cells is a two-step process. First, a development blockade is triggered, followed by a physiological blockade, resulting in a nutrient distributor that is similar to a transfer cell whose task is to feed the nematode (Bird, 1996) (Figure 5).

Recent transcriptome analyses suggest that nutrition cells are different from other plant cells (Gheysen e Mitchum, 2011). Transcriptomes are known to change the plant cell using effector proteins.

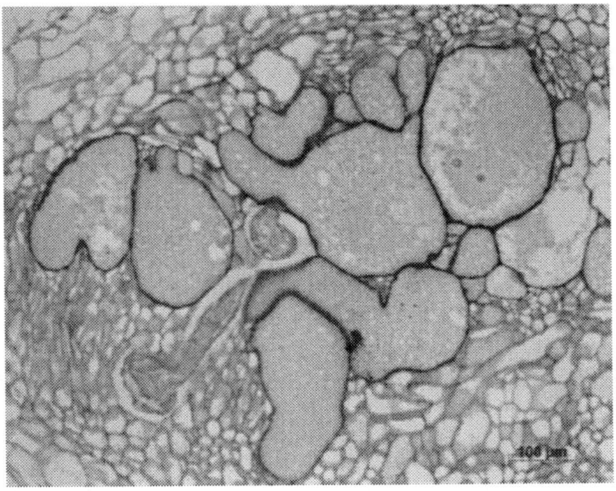

Figure 5. Giant cells (nutrition center) developed by gall-forming nematodes (*M. incognita*) in tobacco (*Nicotiniana tabacum*). (Photos kindly provided by Dr. José Dijair Antonino Jr.).

Parasitic strategies vary across nematode species that colonize plants, from those that only perforate and ingest the cell's contents as they migrate between host cells, to those that induce complex changes in plant cells selected as a permanent feeding site during the sedentary stages of a nematode's lifecycle (Hussey & Grundler, 1998). Phytonematodes are able to mimic plant peptides in order to establish a more effective parasitism, such as CLEs. The proteins are involved in the regulation of plant growth and development aspects.

Nematode effectors are proteins and other molecules secreted by these parasites so as to promote penetration and migration inside the host plant's roots. These molecules also help evade or neutralize the plant's immune response, which in turn start or maintains the establishment of feeding sites based on the changes effected in the host's metabolism. Most effectors are produced in the pharynx glandules (in dorsal and subventral locations), and are delivered to cytoplasm by the stylet, a lancet-like structure in the oral cavity (Davis et al., 2008).

The proteins in the dorsal glandule are associated with the nematode migration process and facilitation at entry. They are formed by a variety of enzymes that degrade cellulose, hemicelluloses or pectins (Abad et al., 2008) and the proteins that bind to components of the cell wall, allowing or speeding up the digestion process, as cellulose binding proteins (Hewezi et al., 2008) and expansin (Qin et al., 2004) do. Glandule proteins change the plant's response to parasitism, in a process in which galls are formed and the parasite establishes itself (Figure 6).

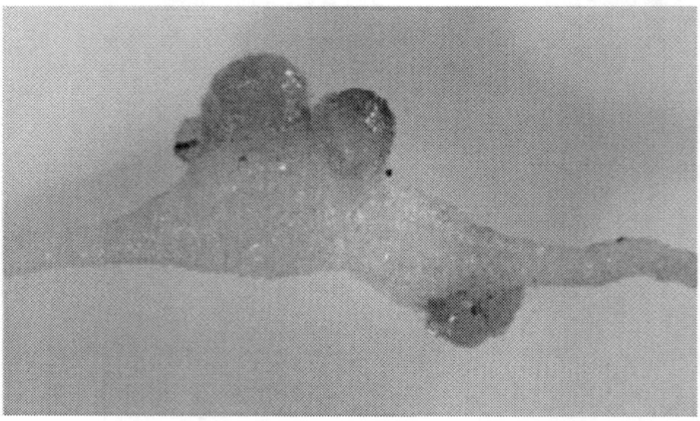

Figure 6. Gall produced in tobacco root (*Nicotiniana tabacum*), with egg masses of *M. incognita* stained with floxin B. (Photos kindly provided by Dr. José Dijair Antonino Jr.).

Their effector proteins influence the regulation of auxin, a plant hormone responsible for organogenesis. The effector protein Hs19C07, of the cyst-producing nematode *Heterodera schachtii* was discovered to interact with AUX3 (LAX3), and its super-expression increased the rate of emergence of lateral roots, indicating a more intense flux of auxin. It has been proposed that Hs19C07 increases LAX3 activity by binding to it. LAX3 is expressed in the development of nutrition cells, and is known to upregulate enzymes that degrade the cell wall (Lee et al., 2011).

Thousands of plant species are known hosts to nematodes; however, some significant exceptions have been described. There are those plants that have evolved to present protective mechanisms that prevent the attraction, lodging, migration as well as the establishment of feeding sites, nutrition, reproduction and the survival of nematodes. Generally speaking, these plants rank under the category of antagonistic species. In principle, antagonistic plants are those that secrete nematicide compounds in the soil that help reduce nematode populations (Embrapa, 2007). But there are also trap plants, which exhibit the interesting ability to kill nematodes, exposing their own roots as baits to these parasites. They attract infectious larvae but do not allow the establishment of the parasitic relationship, which eventually stalls the nematode's lifecycle (Embrapa, 2007). Yet another interesting example of nematode control capability is given by non-host plants, which release repellent substances so as to ward off nematodes (Embrapa, 2007).

In turn, resistant plant species are infected by nematodes, but the parasitic relationship is not established, due to these plant's sophisticated defense mechanisms. Nevertheless, for this trait to manifest, the plant species needs to have resistance genes that trigger a response to the non-virulent pathogen of the parasite, which acts as a signal. In this sense, the plant may develop biochemical defense mechanisms that are commonly associated with the synthesis of secondary metabolites or to the presence of resistance genes. The end result is an incompatible plant-nematode interaction (Embrapa, 2007).

At this point, it is interesting to further the analysis of the response of plants to nematode parasitism. This response, concerning the host's susceptible or resistant genotypes takes place in the feeding sites that surround the nematode's head (Jones, 1981; Williamson & Kumar, 2006), suggesting that the effectors are secreted at a point where the stylet and amphyds (the sensory labial neuron) open an orifice. The analysis of nervous processes and of neurosecretor cells that superimpose cytoplasm extensions of glandular cells and the ampoule also indicate that effector secretion is probably regulated by the nematode nervous system (Endo, 1984; Hussey & Mims, 1990).

The generation of nutrition and resistance cells in the host as part of an active response (hypersensitivity) to phytoparasitic nematodes takes place exclusively around this parasite's head (Davis et al., 2008; Williamson & Kumar, 2006), suggesting that the primary secretions for parasitism originated in the nematode's orifice (Davis et al., 2008; Hussey, 1989) and amphyds (Semblat et al., 2001).

A study by Zhao et al. (2000) revealed that the presence of high concentrations of attractive substances may lead to a null effect, inhibiting the movement of J2 of *M. incognita* aggregated near roots. It is possible to observe that the activity of attractive compounds depend on the balance between the attraction and the repellent stimuli. However, attractive substances released by the host plant root establish a concentration gradient in the soil, directing nematode migration towards the root.

In a study carried out by Qingren Wang et al. (2006), okra (*Albemoschus esculentus* L.) was grown in separate pots containing *Crotolaria juncea* and *Sorghum bicolor* to assess the lesions caused by gall-producing nematodes. The positive results show that there is an antagonistic interaction between these species and the nematodes, and that *C. juncea* is more effective, also improving okra yields. Such combined practices have shown a significant potential in terms of organic agricultural applications, when one plant alone does is not able to fight parasitism.

Nematodes that feed on plants may profit by the adoption of intensive monoculture farming practices (Pattison et al., 2008). A soil exposed to stresses of different kinds or to environmental disturbances, such as intensive farming sites, are places where phytonematodes may thrive (Cares & Huang 1991; Niles &Freckman 1998; Gomes et al., 2003; Mondino et al., 2006).

Plant Lectins

The main characteristic of plant lectins is its ability to bind do specific carbohydrate sites. Plants that synthesize lectins exhibit resistance to nematodes, mainly leguminous species, in which lectins are present at higher concentrations. Plant lectins, mainly those produced by species of the *Leguminosa* family, are considered a model in protein-glycide recognition events. This is due to the fact that they are easily purified and to the wide specificity for glycide residues, even though when these exhibit considerable sequence conservation (Lis & Sharon, 1990), as ConA, ConM and soybean agglutinin (Figure 7).

Figure 7. *Canavalia maritima* (ConM, Leguminosae). a) ConM in its natural habitat. b) ConM and its pod, seeds and flower, which are lectin sources.

The quaternary structure of lectins may also contribute to the recognition of carbohydrates. The ability that each subunit exhibits to individually bind to carbohydrates may develop a unique network of cross-binding structures. This allows lectins to establish complex interactions with cell and matrix surfaces that contain glycoconjugates and multiple binding sites, which probably enhances the biological activity of plant lectins (Loris et al., 1998; Wright, 1997; Loris et al., 1996).

Lectins normally exhibit specificity for glycose, N-acetylglycosamine, and mannose, though some lectins recognize galactose, N-acetylgalactosamine, fucose and oligosaccharides. This binding ability is attributed to sialic acid residues, a family of saccharides that are not synthesized by plants (Reuter, 1996) (Table 2).

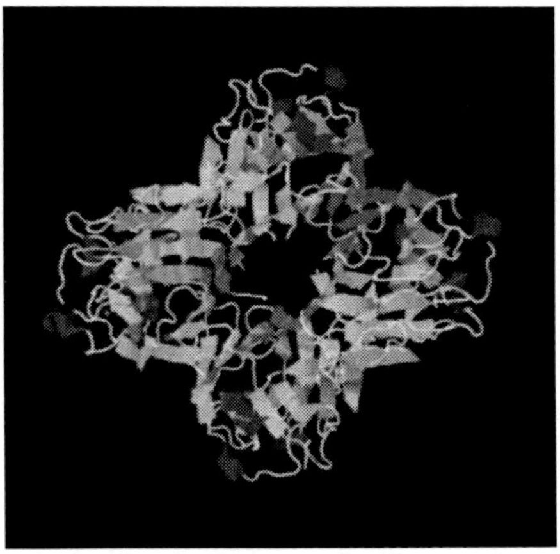

Figure 8. Tetramer-shaped molecular structure of *Canavalia maritima* (ConM).

Table 2. Lectins extracted from some plant species and their specific carbohydrates

Plant species	Specific carbohydrates	
	Monosaccharides	Glycans
Canavalia ensiformis	Mannose/glucose	GlcNAc/β2Manα6(GlcNAcβ2Manα3) Manβ4GlcNAc
Bauhinia purpurea	N-acetylgalactosamine	Galβ3GalNAc
Phaseolus lunatus	N-acetylgalactosamine	GalNAcα3(Fucα2)Galβ-R
Laburnum alpinum	N-acetylglucosamine	GlcNAcβ4GlcNAc
Ulex europaeus I	Fucose	Fucα6GlcNAc
Caragana arborescens	N-acetylglucosamine	GalNAcα3GalNAcβ3Galα4Galβ4Glc

Some plant lectins are toxic to several plant pathogens, which may play a significant role inside seeds and other peripheral plant tissues. In some regions of the world, the transgenic expression of the gene is used to induce the production of a given lectin so as to fight one parasite, such as the agglutinin of *Galanthus nivalis* used to control sap-sucking insect species that infest rice paddy fields (Van Rhijn et al., 1998; Etzler, 1992).

Plant lectins are more than a player in the defense system. As far as plant lectins are concerned, the interaction of these substances with cell surface compounds is also known to have a positive effect: the establishment of new, desirable contacts. Several plants, particularly *Leguminosae* species, are known for their capacity to establish symbiosis with soil bacteria of the *Rhizobium* genus and related genera, which are atmospheric nitrogen fixers, reducing the dependence on external nitrogen sources, like fertilizers (Rüdiger, 1984). Lectins mediate the plant-bacterium symbiosis, similarly to what they do with EPNs.

The search for medicinal plants to control nematodes is the rationale behind several research efforts endeavored to replace synthetic chemicals. However, these plants will only be useful if the compounds they contain meet the same efficacy, safety and quality control criteria as their synthetic counterparts (Rates, 2001).

Essays to Analyze Plant Lectins in Nematodes

In an effort to understand the dynamic of the interaction between the protein and nematodes and to open a viable, efficient and ecologically correct pathway in the control of nematode populations, the nematicidal activity of lectins has been the object of emphatic, consistent research across the globe. These investigations have produced promising results, to the extent that they afford to more fully understand the glycolysation process and the way some plants evolve defense mechanisms.

A study carried out by Santi-Gadelha et al. (2012) has demonstrated that the lectin extracted from *Acacia farnesiana* (ACAL) exhibits an interesting nematicidal potential against *Meloidogyne incognita*, both at egg hatching and against juveniles (J2) which varies with exposure times and therefore may be used in vermicide formulations, introduced in the composition of natural fertilizers or, once sequenced, be integrated in the transgenic expression process of plant that are affected by gall-producing nematodes..

The nematode *Bursaphelenc husxylophilus* is transmitted by *Monachamus alternates* H. Lectins produced by the species *Pinelia ternate* (PTA agglutinin) and *Lycoris radiata* (LRA agglutinin) exhibited significant nematicidal action. LRA was found to reduce nematode populations by over 50% within a 96-h period (Qi Gaofu et al., 2008). It has also been shown that the higher the concentration of lectins, the greater the toxic effects on nematodes.

Several lectins that exhibit biological activity have been tested against phytonematodes in comparative studies. Zhao et al. (2009) tested antifungal lectins and proteins against the nematodes *Ditylenchus dipsaci* and *Heterodera glycines*, observing that the lectins produced by the fungi species *Xylaria hypoxylon, Agrocybe cylindracea* and *Tricholoma mongolicum (TML-1)* are more potently active against *D. dipsaci*. The lectins produced by *Pseudostellaria heterophylla, Samta tomate*, the mushrooms *T. mongolicum* (TML-2), *Ganoderma lucidum* and *Boletus edulis*, and antifungal proteins present in *Ginkgo biloba* and the pumpkin *Cucurbita moschata* have antinematode action that are considerably lower against *D. dipsaci*, and are not considered feasible control alternatives. However, the authors report that all these lectins were significantly effective against *Heterodera glycines*.

Spiegel et al. (1996) observed that, when *M. incognita* was exposed to *Canavalia eosiformis* in combination with the agglutamin present in wheat germ, the lodging capacity of the parasite infecting the host was not the same as that exhibited by its cuticle to the endospores of the host. This was proved when J2 was treated with carbohydrates (fucose ou α-methyl manoside, N-acetyl-D-glycosamine, D-xylose), and the results were the same.

Final Comments

It remains to be established how lectins affect the release of effector proteins. When these glycoconjugates bind to the carbohydrates present in chemoreceptors that bind to sensory cells, where the stimulus reaches the amphyd, the likely result may be a change that could lead to lower rates of effector proteins and mimetic peptides such as CLE, since the release of these compounds is mediated by the nervous system, such as described in the literature, negatively affecting lodging capacity and maintenance of parasitism.

Lectins may be used as nematicides. These molecules may bind to carbohydrates in the whole nematode, exerting cytotoxic effects and killing the parasite, with no harm to the environment. Another action mechanism exhibited by lectins is the modulation activity, since all carbohydrates in the membrane of the nematode block the action of the host's immune system. Lectins would also block sugars, promoting the role of antibodies, thus eliminating the nematode.

The incorporation of genes expressing lectins with specific nematicidal action may be the solution to be adopted by small farmers in northeastern Brazil. The large volume of pesticides used by these farmers contaminates soils and water bodies, affecting the nutritional value of products as commonly associated with intense monoculture farming, apart from contributing to the desertification process, already evident.

Acknowledgments

The authors thank Dr José Dijair A. de Souza Jr, Laboratório de Interação Molecular Planta-Praga, Embrapa Cenargen Brasília/DF; Dr Carlos Eduardo Rossi, Nematologist, Instituto Agronômico de Campinas/SP; Dr Carlos Eduardo Winter, Departamento de Parasitologia/ICB/USP, São Paulo/SP.

References

Abad, P; Favery, B; Rosso, MN; Castagnone-Sereno, P. Root-knot nematode parasitism and host response: molecular basis of a sophisticated interaction. *Mol Plant Pathol*, 2003, 4, 217-224.

Abad, P; Gouzy, J; Aury, JM; Castagnone-Sereno, P; Danchin, EGJ; Deleury, E; Perfus-Barbeoch, L; Anthouard, V; Artiguenave, F; Blok, VV; Caillaud, MC; Coutinho, PM; Dasilva, C; De Luca, F; Deau, F; Esquibet, M; Flutre, T; Goldstone, JV; Hamamouch, N; Hewezi, T; Jaillon, O; Jubin, C; Leonetti, P; Magliano, P; Maier, TR; Markov, GV; McVeigh, P; Pesole, G; Poulain, J; Robinson-Rechavi, M; Sallet, E; Ségurens, B; Steinbach, D; Tytgat, T; Ugarte, E; van Ghelder, C; Veronico, P; Baum, TJ; Blaxter, M; Bleve-Zacheo, T; Davis, EL; Ewbank, JJ; Favery, B; Grenier, E; Henrissat, B; Jones, JT; Laudet, V; Maule, AG; Quesneville, H; Rosso, MR; Schiex, T; Smant, G; Weissenbach J; Wincker, P. Genome sequence of the metazoan plant parasitic nematode Meloidogyne incognita. *Nat Biotechnol*, 2008, 26.doi,10.1038/nbt.1482.

Akhurst, RJ; Dunphy, GB. Interações tripartido entre bactérias simbioticamente associadas, nematóides entomopatogênicos, e seus hospedeiros de insetos. In: Beckage, N; Thompson, S; Federici, B editors *Parasitas e patógenos de insetos vol. 2*, New York: Academic Press, 1993, 1-23.

Bird, DM. Manipulation of host gene expression by root-knot nematodes. *J Parasitol*, 1996, 82, 881-888.

Campos, RA; Boldo, JT; Pimentel, IC; Dalfovo, V; Araújo, WL; Azevedo, JL; Vainstein, MH; Barros, NM. Endophytic and entomopathogenic strains of *Beauveria* sp to control the bovine tick *Rhipicephalus (Boophilus) microplus*. Online Journal *Genetics and Molecular Research*, 2010. DOI 10.4238/vol9-3gmr884.

Cares, JH; Huang, SP. Nematode fauna in natural and cultivated cerrados of Central Brazil. *Fitopatologia Brasileira*, 1991,16,199–209.

Chitwood, DJ. Research on plant-parasitic nematode biology conducted by the United States Department of Agriculture-Agricultural Research Service. *Pest Manag Sci*, 2003,59,748-753.

Chrispeels, MJ; Raikhel, NV. Lectins, lectin genes, and their role in plant defense. *Plant Cell*, 1991, 3,1-9.

Clarke, DJ. *Photorhabdus*: a model for the analysis of pathogenicity and mutualism. *Cell Microbiol*, 2008,10, 2159-2167.

Cummings, RD; Esko, JD. Nematoda. In: Varki A, Cummings RD, Esko JD, Freeze HH, Stanley P, Bertozzi CR, Hart GW, Etzler ME editors. *Essentials of Glycobiology. 2nd edition.* Cold Spring Harbor (NY): Cold Spring Harbor Laboratory Press; 2009; 333-346. Available at http://www.ncbi.nlm.nih.gov/books/NBK1930/

Davis, EL; Hussey, RS; Baum, TJ. Getting to the roots of parasitism by nematodes. *Trends Parasitol*, 2004, 20,134-141.

Davis, EL; Hussey, RS; Mitchum, MG; Baum, TJ. Parasitism proteins in nematode-plant interactions. *Curr Opin Plant Biol*, 2008,11,360-366.

Davis, EL; Hussey, RS; Mitchum, MG; Baum, TJ. Parasitism proteins in nematode-plant interactions. *Curr Opin Plant Biol*, 2008,11, 360–66.

De Almeida-Engler, J; De Vleesschauwer, V; Celenza, JR; Inze, D; Van Montagu, M; Engler, G; Gheysen, G. The use of molecular markers and cell cycle inhibitors to analyze cell cycle progression in nematode-induced galls and syncytia. *Plant Cell*, 1999, 11, 793-807.

Ehlers, RU; Hokkanen, HMT. Insect biocontrol with nonendemic entomopathogenic nematodes (*Steinernema* and *Heterorhabditis* spp): conclusions and recommendations of a combined OECD and COST workshop on scientific and regulatory policy issues. *Biocontrol Sci Technol*, 1990, 6, 295-302.

Endo, BY. Ultrastructure of the esophagus of larvae of the soybean cyst nematode, Heterodera glycines. *Proceedings of the Helminthological Society of Washington*, 1984, 51, 1–24.

Etzler, ME. Plant lectins: Molecular biology, synthesis, and function. In: Allen HJ, Kisailus EC editors *Glycoconjugates: Composition, structure and function.* New York: Marcel Dekker, 1992, 521–539.

Ferraz, LCCB; Monteiro, AR. Nematóides. In: Filho, AB; Kimati, H; Amorim, L editors. *Manual de fitopatologia*. São Paulo: Agronômica Ceres; 1995,168-201.

Freitas, LG; Oliveira, RDL; Ferraz, S. *Introdução a Nematologia*. Viçosa: Editora UFV, 2001.

Gaofu, Q; Shiqing, M; Fayin, Z; Zhiniu, Y; Xiuyun, Z. In vitro assessment of plant lectins with anti-pinwood nematode activity. *J Invertebr Pathol*, 2000, 98, 40–45.

Gheysen, G; Mitchum, MG. How nematodes manipulate plant development pathways for infection. *Curr Opin Plant Biol*, 2011, 144,15-421.

Gheysen, G; Mitchum, MG. How nematodes manipulate plant development pathways for infection. *Curr Opin Plant Biol*, 2011, 14,415–421.

Gomes, GS; Huang, SP; Cares, JE. Nematode community. trophic structure and population fluctuation in soybean fields. *Fitopatologia Brasileira*, 2003, 28, 258–266.

Gupta, VVSR; Yeates, GW. Soil microfauna as bioindicators of soil health. In Pankhurst, CE; Doube, BM; Gupta, VVSR editors. *Biological Indicators of Soil Health*, London: CABI, 1997, 201–233.

Heungens, K; Cowles, CE; Goodrich-Blair, H. Identification of *Xenorhabdus nematophila* genes required for mutualistic colonization of *Steinernema carpocapsae* nematodes. *Mol Microbiol*, 2002, 45, 1337-1353.

Hewezi, T; Howe, P; Maier, TR; Hussey, RS; Mitchum, MG; Davis, EL; Baum, TJ. Cellulose binding protein from the parasitic nematode Heterodera schachtii interacts with Arabidopsis pectin methylesterase: cooperative cell wall modification during parasitism. *Plant Cell*, 2008, 20, 3080-3093.

Hussey, RS. Disease-inducing secretions of plant-parasitic nematodes. *Annu Rev Phytopathol*, 1989, 27, 123–41.

Hussey, RS; Grundler, FMW. Nematode parasitism of plants. In: Perry, RN; Wright DJ editors. *The physiology and biochemistry of free-living and plant parasitic nematodes*. Wallingford, UK: CABI Publishing; 1998, 213–243.

Hussey, RS; Mims, CW. Ultrastructure of esophageal glands and their secretory granules in the root-knot nematode Meloidogyne incognita. *Protoplasma*, 1990, 156, 9–18.

Jones, MGK. Host-cell responses to endo-parasitic nematode attack – structure and function of giant-cells and syncytia. *Ann Appl Biol*, 1981, 97, 353–372.

Jones, MGK. Host-cell responses to endo-parasitic nematode attack: structure and function of giant-cells and syncytia. *Ann Appl Biol*, 1981, 97, 353–72.

Jonsson, HB. Receptors and recognition in nematodes. In: Veech, JA; Dickson, DW editors *Vistas on Nematology*, Hyattsville: Society of Nematology, 1987, 153-158.

Kandji, ST; Ogol, CKPO; Albrecht, A. Diversity of plant-parasitic nematodes and their relationships with some soil physic-chemical characteristics in improved fallows in western Kenya. *Appl Soil Ecol*, 2001, 18, 143–157.

Kimenju, JW; Karanja, NK; Mutua, GK; Rimberia, BM; Wachira, PM. Nematode community structure as influenced by land use and intensity of cultivation. *Tropical and subtropical agroecossystems*, 2009, 11, 353–360.

Lambshead, PJD. Recent developments in marine benthic biodiversity research. *Oceanis*, 1993, 19, 5-24.

Lee, C; Chronis, D; Kenning, C; Peret, B; Hewezi, T; Davis, EL; Baum, TJ; Hussey, RS; Bennett, M; Mitchum, MG. The novel cyst nematode effector protein 19C07 interacts with the Arabidopsis auxin influx transporter LAX3 to control feeding site development. *Plant Physiol*, 2011, 155, 866–880.

Lis, H; Sharon, N. Legume lectins - a large family of homologous proteins, *FASEB J*, 1990, 4, 3198-3208.

Loris, R: Hamelryck, T; Bouckaert, J; Wyns, L. Legume lectin structure. *Biochim Biophys Acta*, 1998, 1383, 9–36. [PubMed: 9546043]);

Loris, R: Maes, D; Poortsmans, F; Wyns, L; Bouckaert, J. A structure of the complex between concanavalin A and methyl-3,6-di-O-(α-D-mannopyranosyl-α-D-mannopyranoside reveals two binding modes. *J Biol Chem*, 1996, 271,30614–30618. [PubMed: 8940035].

Marban-Mendonza, N; Jeyaprakash, A; Janson, HB; Damon, JRA; Zuckerman, BM. Control of root-knot nematodes on tomato by lectins. *J Nematol*, 1987, 19, 331-335.

Medeiros, J; Rosa, JS; Tavares, J; Simões, N. Susceptibility of *Pseudaletia unipuncta* (Lepidoptera: Noctuidae) to entomophatogenic nematodes (Rhabditida: Steinernematidae and Heterorhabditidae) isolated in the Azores: effect of nematode in strain and host age. *J Econom Entomol*, 2000, 93, 1403-1408.

Mondino, EA; Chaves, EJ; Clausen, AM. Distribution of nematodes in potato fields soil in Andean Valleys of Argentina. *Revista Facultad de Agronomia UBA*, 2006, 26, 141–148.

Mondino, EA; Tavares, OCH; Ebeling, AG; Figueira, AF; Quintero, EI; Berbara, RLL. Avaliação das comunidades de nematóides do solo em agroecossistemas orgânicos. *Acta Scientiarum Agronomy*, 2009, 31, 509–515. *On-line version* ISSN 1807-8621 http://dx.doi.org/10.4025/ actasciagron. v31i3.433

Niles, RK; Freckman, DW. From the ground up: nematode ecology in bio-assessment and ecosystem health. In: Bartles, JM editor. *Plant and nematode interactions.* Madison: ASA/CSSA/SSSA, 1988, 65–85.

Ortiz-Urquiza, A; Riveiro-Miranda, L; Santiago-Álvarez, C; Quesada-Moraga, E. Insect-toxic secreted proteins and virulence of the entomopathogenic fungus *Beauveria bassiana. J Invertebr Pathol*, 2010, 105, 270-280.

Pattison, AB; Moody, PW; Badcock, KA; Smith, LJ; Armour, JA; Rasiah, V; Cobon, JA; Gulino, L-M; Mayer, R. (2008). Development of key soil health indicators for the Australian banana industry. *Appl Soil Ecol*, 40,155–164.

Perry, RN; Aumann, J. Behaviour and sensory responses. In: Perry, RN; Wright, DJ editors. *The physiology and biochemistry of free-living and plant-parasitic nematodes*. Wallingford: CAB International, 1998, 75-269.

Peumans, WJ; Van Damme, EJ. Lectins as plant defense proteins. *Plant Physiol*, 1995, 109, 347-352.

Poinar, GO. *Nematodes for biological control of insects*. Boca Raton, USA: CRC Press; 1979.

Popiel, I; Hominick, WM. Nematodes as biological control agents: part II. *Adv Parasitol*, 1990, 31, 381-431.

Qi, Y; Hu, C. Soil nematode abundance in relation to diversity in different farming management system. *World Journal of Agricultural Sciences*, 2007, 3, 587–592. ISSN: 1817-3047. IDOSI Publications.

Qingren, W; Yuncong, L; Waldemar, K; Zafar, H. Influence of cover crops and soil amendments on okra (Abelmoschus esculentus L.) production and soil nematodes. *Renewable Agriculture and Food Systems*, 2006, 22, 41–53. doi:10.1017/S1742170507001585

Rates, SMK. Plants as source of drugs. *Toxicon*, 2001,39,603-13.

Reuter, G; Gabius, HJ. Review: Sialic acids. Structure, analysis, metabolism, and recognition. *Biological Chemistry Hoppe-Seyler*, 1996, 377, 325–342. ISSN: 0177-3593.

Rocha Fragoso, R; Tristan Lourenço, I; Barbosa Viana, AA; dos Santos de Lima Souza, D; Vieira de Andrade, R; Mehta, A; Miranda Brasileiro, AC; de Campos Pinto, ER; a de Lima, LM; Lima Rocha, T; Grossi de As. MF. *Documentos 198 - Interação Molecular Planta-Nematóide*, Planaltina, DF : Embrapa Cerrados, 2007. ISSN 1517-5111.

Rüdiger, H. On the physiological role of plant lectins. *BioScience*, 1984, 34, 95–99.

Santi-Gadelha, T; Rocha, BAM; Gadelha, CAA; Silva, HC; Castellon, RER; Gonçalves, FJT; Toyama, DO; Toyama, MH; de Souza, AJF; Beriam, LOS; Martins, JL; Joazeiro, PP; Cavada, BS. Effects of a lectin-like protein isolated from Acacia farnesiana seeds on phytopathogenic bacterial strains and root-knot nematode. *Pestic Biochem Physiol*, 2012, 103, 15-22.

Semblat, JP; Rosso, MN; Hussey, RS; Abad, P. Castagnone-Sereno, P. Molecular cloning of a cDNA encoding an amphid-secreted putative avirulence protein from the root-knot nematode *Meloidogyne incognita*. *Mol Plant-Microbe Interact*, 2001, 14, 72–79.

Spiegel, Y; Mor, M; Sharon, E. Attachment of *Pasteuria penetrans* endospores to the surface of *Meloidogyne javanica* second-stage juveniles. *J Nematol Lakeland*, 1996, 28, 328-334.

Toledo, AV; de Remes Lenicov, AM; Lopez-Lastra, CC. Histopathology caused by the entomopathogenic fungi, *Beauveria bassiana* and *Metarhizium anisopliae*, in the adult planthopper, *Peregrinus maidis*, a maize virus vector. *J Insect Science*, 2010, 10, 35.

Trudgill, DL; Blok, VC. Apomictic, polyphagous root-knot nematodes: exceptionally successful and damaging biotrophic root pathogens. *Annu Rev Phytopathol*, 2001, 39, 53-57.

Van Rhijn, P; Goldberg, RB; Hirsch, AM. *Lotus corniculatus* nodulation specificity is changed by the presence of a soybean lectin gene. *Plant Cell*, 1998, 10, 1233–1250. [PMC free article: PMC144063] [PubMed: 9707526]]

Vieira, PC; Mafezoli, J; Biavatti, MW. Inseticidas de origem vegetal. In: Ferreira, JTB; Corrêa, AG; Vieira, PC editors *Produtos naturais no controle de insetos.* São Carlos: EdUFSCar, 2001, 23-45

Walla, RK; Nandal, SN; BHATTI, D S. Nematicidal efficacy of plant leaves and *Paecillomyces lilacinus*, alone or in combination, in controlling *Meloydogyne incognita* on okra and tomato. *Nematologica Mediterranea*, 1999, 27, 3-8.

West, RJ; Vrain, TC. Nematode control of black army cutworm (Lepidoptera: Noctuidae) under laboratory and field conditions. *Can Entomol*, 1997, 129, 229-239.

Williamson, C. A Gleason. Plant-nematode interactions. *Curr Opin Plant Biol*, 2003, 6, 327-333.

Williamson, VM; Kumar, A. Nematode resistance in plants: the battle underground. *Trends Genet*, 2006, 22, 396-403.

Williamson, VM; Kumar, A. Nematode resistance in plants: the battle underground. *Trends Genet*, 2006, 2, 396–403.

Williamson, VM; Kumar, A. Nematode resistance in plants: the battle underground. *Trends Genet*, 2006, 22, 396–403.

Wood, WB. The Nematode *Caenorhabditis elegans*. Cold Spring Harbor: Cold Spring Harbor Laboratory Press, New York, 1988.

Wright, CS. New folds of plant lectins. *Curr Opin Struct Biol*, 1997, 7, 631–636. [PubMed: 9345620].

Wright, DJ. Nematicides: mode of action and new approaches to chemical control. In: Zuckerman, BM; Rohde, RA editors. *Plant parasitic nematodes*. New York: Academic Press, 1981, 421-449.

Yeates, GW; Bongers, T. Nematode diversity in agroecosystems. *Agriculture, Ecosystems and Environment* 1999,74,113–135. ISSN 0167-8809; http://dx.doi.org/10.1016/S0167-8809 (99) 00033-X

Yeates, GW; Pattison, AB. Moving up the Food Chain: Protozoa and nematodes. In Uphoff, N; Ball, AS; Fernandes, E; Herren, H; Husson, O; Laing, M editors. *Biological approaches to sustainable soil systems*, New York: CRC/Marcel Dekker, 2006, 149–162.

Zhao, S; Guo, XY; Liu, QH; Wang, HX; Ng, TB. Lectins but not antifungal proteins exhibit anti-nematode activity. *Environmental Toxicology and Pharmacology*, 2009, 28, 265-268.

Zhao, X; Schmitt, M; Hawes, CM. Species-dependent effects of border cell and root tip exudates on nematodes behavior. *Phytopathology, St. Paul* 2000;90:1239-1245.

Zuckerman, BM. Hypotheses and possibilities of intervention in nematode chemoresponses. *J Nematol Lakeland*, 1983,15,173-182.

Zuckerman, BM; Jansson, HB. Nematode chemotaxis and possible mechanisms of host/prey recognition. *Annu Rev Phytopathol Palo Alto*, 1984, 22, 95-113.

In: Nematodes
Editor: Lee M. Davis
ISBN: 978-1-62948-764-9

Chapter 3

SOME ASPECTS OF BIODIVERSITY AND DISTRIBUTION OF NEMATODES BELONGING TO THE FAMILIES *LONGIDORIDAE* AND *TRICHODORIDAE*

***N. D. Romanenko*[1*] *and V. Krivtsov*[2,3†]**
[1] Center of Parasitology of A.N. Severtsov Institute of Ecology and Evolution , Russian Academy of Sciences, Moscow, Russia
[2]CECS, The Crew Building, King's Buildings, University of Edinburgh, West Mains Road, Edinburgh, Scotland, UK
[3]Department of Ecology, Kharkov State University, Kharkov, Ukraine

ABSTRACT

This article presents information related to the biodiversity and distribution of nematodes belonging to the families *Longidoridae* and *Trichodoridae* in different habitats in the countries belonging to the CIS (Commonwealth of Independent States, formerly known as the USSR).The investigations have been conducted in 13 republics of the former USSR, and in 83 regions of the Russian Federation. The studies revealed 40 species of *Longidoridae,* including 18 species belonging to the genus *Longidorus,* 3 belonging to the genus *Paralongidorus*, and 19

* Tel.: (495)9523145, Fax: (495)9523146, E-mail: cenologypathlab@mail.ru.
† Email: e96kri69@netscape.net.

belonging to *Xiphinema*. The list of species belonging to the family *Trichodoridae* is represented by 12 species, including 9 species belonging to the genus *Trichodorus,* and 3 belonging to *Paratrichodorus.* Most of these nematodes are characteristic for the palearctic fauna. The combination of field and laboratory studies allowed us to assess the importance of various agroecological factors (e.g. host plants, geographical distribution, location, climatic conditions, edaphic characteristics/soil properties, etc.) for population abundance and the occurrence of particular species.

Keywords: Nematodes, biodiversity, geographical distribution, agroecological factors, population, host plants

INTRODUCTION

This article presents the summary of long-term investigations and literature information related to the biodiversity and distribution of nematodes belonging to the families *Longidoridae* and *Trichodoridae* in different habitats in the countries belonging to the CIS (Commonwealth of Independent States, formerly known as the USSR). The investigations have been conducted in 13 republics of the former USSR, and in 83 regions of the Russian Federation. The combination of field and laboratory studies allowed us to assess the importance of various agroecological factors (e.g. host plants, geographical location, climatic conditions, edaphic characteristics/soil properties, etc.) for population abundance and the occurrence of particular species.

RESULTS AND DISCUSSION

The summary of the nematodes' occurrence on specific plant species is given in Table 1. The studies revealed 40 species of *Longidoridae,* including 18 species belonging to the genus *Longidorus,* 3 belonging to the genus *Paralongidorus*, and 19 belonging to *Xiphinema*. The list of species belonging to the family *Trichodoridae* is represented by 12 species, including 9 species belonging to the genus *Trichodorus,* and 3 belonging to *Paratrichodorus.* A number of viruses has been identified, and 15 species (belonging to *Longidoridae* and *Trichodoridae*) have been revealed to transmit these viruses.

As has previously been shown (Romanenko 1993, 1994, 2004, 2006), the genera *Longidorus* and *Xiphinema* both have a wide distribution. The sizes of their populations depend on host plants, soil types, etc. The most abundant among the genus *Xiphinema* are *Xiphinema taylori* (probably a synonym of *Xiphinema brevicolle*), *Xiphinema diversicaudatum*, *Xiphinema pachtaicum* and *Xiphinema paramonovi.* These species are found in all the regions of intensive agriculture. In the southern areas (Central-chernozem zone, southern Russia and Ukraine, Caucasus and middle Asia), the most common species are *Xiphinema index, Xiphinema vuittenezi* and *Xiphinema turcicum.*

Representatives of the genus *Longidorus* are also widely distributed. In the middle of the European part of Russia the commonest species are *Longidorus elongatus, Longidorus attenuatus* and *Longidorus leptocephalus.* In the southern part of the former USSR the commonest species are *Longidorus africanus, Longidorus martini* and *Longidorus sylphus.*

The genus *Paralongidorus* is represented only by three species. The species *Paralongidorus maximus and Paralongidorus nudus* are found in various edapho-climatic zones. The species *Paralongidorus georgiensis* was found only in the republic of Georgia.

The distribution of particular species within the family *Longidoridae* depends on complex relationships among biological and abiotic factors, including host plants, macro and micro relief, edapho-climatic characteristics, soil mineral content and anthropogenic influences (Romanenko 1993, 1994, 2004, 2006). Nematodes belonging to the family *Longidoridae* are encountered both on wild and cultivated plants. It has been suggested that the process of transformation from wild to cultivated hosts has been active since the incidence of agricultural practices, and continues at present (Romanenko, 1985, 1993).

Lab experiments have demonstrated trophic preference of the most common species. In particular, the favoured host plant for *X. index* and *X. vuittenezi* was vine *Vitis vinifera*; for *L. elongatus – Fragaria vesca and F. ananassa, Lolium perenne* and sugar beet (*Beta vulgaris)*; for *X. taylori* –apple and pear trees, netles; for *X. pachtaicum – Chenopodium quinoa*; *X. diversicaudatum - Fragaria ananassa, F.vesca*; *X. paramonovi* – white clover (*Trifolium alba*), celery (*Apium graveolens*), and of black currant (*Ribes nigrum*). Observations in the field revealed that the highest occurrence and population abundance of *X. index* and *X. vuittenezi* is associated with vine; *L. elongatus* and *X. diversicaudatum* with *Fragaria vesca*; *X. taylori* – with nettles, and apple and pear trees; whilst such southern species as *L. africanus, L. sylphus, L. tardicauda, X. index* and *X. vuittenezi* are mainly found in the rhizosphere of vine and some other cultivated subtropical agricultural plants.

Other species (e.g. *L. elongatus X. taylori, X. pachtaicum*) have a wide range of plant hosts (Romanenko 1993, 1994, 2004, 2006).

As regards climatic preferences, the most cold tolerant species are *L. elongatus, L. attenuatus* and *X. pachtaicum*, who are most abundant in the North, North West, and central regions of Russia, as well as in West Siberia and Far East. All these areas are characterised by cold and very continental climate with very sever cold winters, with temperatures below -45 degrees. On the other hand, heat tolerant species inhabit southern areas with subtropical and continental climate in the lowland and mountainous regions of areas of the ancient mediterranean (*L. africanus, L. lignosus, L. macrosoma, L. sylphus, L. tardicauda, L. vineacola, X. attorodorum, X. index, X. vuittenezi, X. riparia, X. turcicum, X. ingens, etc.* A number of species (e.g. *L. elongatus, X. taylori, X. pachtaicum, X. diversicaudatum*) are euriadaptive, and are found in big numbers in the conditions of subtropical, temperate and cold climate.

A number of species favour wet conditions of wetlands and bogs, floodplain forests and flushes with ground water seepage. Such species include, e.g., *L. elongatus, L. leptocephalus, X. diversicaudatum, and X. pachtaicum*. These nematodes are mainly found close to rivers, and have been observed in the proximity of the channels of Lichoborka, Volga, Klyazma, Dniepr, Ob', and Amur, among others. On the other hand, drought tolerant species include *L. africanus, L. lignosus, L. macrosoma, L. sylphus, L. vineacola, L. attorodorum, X. ingens, X. index, X. riparia, X. turcicum, and X. vuittenezi*. These species are found in the steppe, semidesert, and mountainous regions with deep positioning of the ground water table.

It is known, that despite a wide distribution nematodes are adapted to relatively narrow ranges of pH. In the former USSR the nematodes belonging to the family *Longidoridae* occur in the pH range of 3.6 – 8. However, more than 60% occur in soils with pH of 4.6 – 6.5. The majority favour soils with light granulometric content, although *X. diversicaudatum and L. macrosoma* appear to prefer heavy soils with high clay and silty clay content. As regards nematodes preference for soil nutrient content, the preferred ranges for N, P and K are, respectively, 8.31-21.2, 0.25-30.5, and 5-22 mg/100g soil dwt.

Most of the nematodes referred to in this article are representatives of the Palearctic zone, which is characteristic for Russia. Some of the species, however, may be characterised as representatives of other zoogeographic regions/subregions, e.g. mediterranean (*X. italiae, X. distinctus, X. vineacola, X. turcicum*), East American (species of group *X. americanum)*, Indo-Malaysian (*X. elongatum*), Ethiopian (*X. attorodorum*), and neotropical (*X.*

brevicolle and other species with similar ecological and morphological characteristics).

Table 1. The list of nematodes belonging to families Longidoridae (Nematoda: Dorylaimida) and Trichodoridae (Nematoda: Triplonchida), encountered in the former USSR (CIS). Species names are giving according to the publications of Andrassy (2009), Decraemer (1995 , 1998) and Romanenko (1993, 1994, 2006)

Latin names of nematode species	Host plants with which particular species were found in association (see legend for details)														
	A	B	C	D	E	F	G	H	I	J	K	L	M	N	O
Longidorus elongatus	+	+	+	+	+	+	+	+	+	+	+	+	+	+	+
L. africanus	-	-	-	-	-	-	+	-	-	-	-	-	-	-	-
L. arfamerienosus	-	-	-	-	-	-	-	-	-	-	-	-	-	-	+
L. attenuatus	+	+	+	+	+	+	-	-	-	-	-	-	-	-	+
L. caespiticola	+	-	-	-	-	-	-	-	-	-	-	-	-	-	-
L. distinctus	+	-	-	-	+	-	-	-	-	-	-	-	-	-	+
L. euonymus	-	-	-	-	-	-	-	-	-	-	-	+	+	-	-
L. leptocephalus	+	-	-	+	-	-	-	-	-	-	-	-	-	-	-
L. lignosus	-	-	-	-	-	-	-	-	-	+	-	-	-	-	+
L. macrosoma	-	-	-	+	+	+	-	-	-	-	-	-	-	-	+
L martini	-	-	-	-	-	-	+	-	-	-	-	-	-	-	-
L. olegi	-	-	-	-	+	-	-	-	-	-	-	-	-	-	-
L. profundorum	+	-	-	-	-	-	-	-	-	-	-	-	-	-	-
L. rubi	+	-	-	-	+	-	-	-	-	-	-	-	-	-	-
L. striola	-	-	-	-	-	-	-	-	-	-	-	-	-	-	+
L. sylphus	-	-	-	-	-	-	+	-	-	-	-	-	-	-	-
L. tardicauda	-	-	-	-	+	-	+	-	-	-	-	-	-	-	+
L. vineacola	-	-	-	-	-	+	-	-	-	-	-	-	-	-	+
Paralongidorus maximus	-	-	-	-	+	-	-	-	-	-	+	-	-	-	-
P. georgiensis	-	-	-	-	-	-	-	+	-	-	-	-	-	-	-
P. nudus	-	-	-	-	-	-	-	-	-	-	-	-	-	-	+
Xiphinema americanum sl	+	-	+	+	+	+	+	-	+	+	+	+	+	+	+
X. artemisiae	-	-	-	-	-	-	-	-	-	-	-	-	-	-	+
X. attorodorum	-	-	-	-	+	-	-	-	-	-	-	-	-	-	-
X. basilgoodeyi	-	-	-	+	-	-	-	-	+	-	-	-	-	-	-
X. brevicolle	+	+	+	+	+	+	+	+	+	+-	+	+	+	+	+
X. diversicaudatum	+	+	+	+	+	+	+	-	-	+	+	+	+	+	+
X. elongatum	-	-	-	-	-	-	-	-	-	-	-	-	-	-	+
X. ingens	-	-	-	-	-	-	-	-	-	+	+	-	-	-	-

Table 1. (Continued)

Latin names of nematode species	Host plants with which particular species were found in association (see legend for details)														
	A	B	C	D	E	F	G	H	I	J	K	L	M	N	O
X. incognitum	-	-	-	-	-	-	-	-	-	-	-	-	-	-	+
X. index	+	+	+	+	+	+	+	-	-	-	-	-	-	-	+
X. italiae	-	-	+	-	+	-	-	-	-	-	-	-	-	-	-
X. opisthohysterum	-	-	-	-	-	-	-	-	-	-	-	-	-	-	+
X. pachtaicum	+	+	+	+	+	-	+	-	-	-	-	-	-	-	+
X. paramonovi	+	+	-	+	+	+	-	-	-	-	-	-	-	-	-
X. riversi	-	-	-	+	-	-	-	-	-	-	-	-	-	-	-
X. riparia	-		-	-	-	-	-	-	-	-	+	-	-	-	+
X. rotundatum	-	-	-	-	-	-	-	-	-	-	-	-	-	-	+
X. turcicum	-	-	-	-	-	-	-	-	-	-	-	-	-	-	+
X. vuittenezi	+	+	+	-	-	-	+	-	-	-	-	-	-	-	+
**Trichodorus altaicus*															
T. mirabilis	-	-	-	-	-	-	-	-	-	-	-	-	-	-	+
**T. obscurus*															
**T. cylindricus*															
Trichodorus primitivus	-	-	-	-	-	-	-	+	-	-	-	-	-	-	+
T. similis	-	-	-	-	-	-	-	+	-	-	-	-	-	-	+
**T. sparsus*															
**T. viruliferus*															
T. yokooi	-	-	-	-	-	-	-	-	-	-	-	-	-	-	+
Paratrichodorus pachydermus	+	+	+	+	+	+	-	+	-	-	-	+	-	-	+
P. teres	-	-	-	-	-	-	-	+	-	-	-	-	-	-	+
**P. minor* (syn. *P. christei)*															

Note: - absence of the species on the mentioned plant; + presence of the species on the mentioned plant; * the information on host plants needs further research and/or verification.

A -*Malus spp.,(M.domestica, M.silvestris*).

B- *Pyrus spp.,(P.communis, P.domestica, P.caucasica*).

C -*Cerasus sp.,(C.vulgaris)*, *Prunus spp.,(P.domestica*).

D -*Fragaria spp.,(Fragaria* x *ananassa, F. vesca, F. alpina).*

E –*Rubus spp., (R.idaeus*).

F - *Ribes spp., (R.nigrum, R.sibirica, R.alpinum).*

G -*Vitis vinifera*.

H - Solanum tuberosum, S.lycopersicum (Lycoperscon esculentum).

I - *Rosa canina, R.rugosa, Crataegus sanguinea.*

J - *Acer spp.,(A.tataricum), Populus spp., (P. tremula, P. deltoides); Quercus spp.,(Q.robur), Betula verrucosa.*

K - *Hordeum spp., (H.vulgare, H.distichum), Avena sativa, Secale cereale, Triticum aestivum.*
L - *Poa pratensis, P. trivialis, Dactylis glomerata; Trifolium pratense, T. rubens, T. repens, T. alba.*
M - *Salix fragilis, Carex spp.*
N - *Urtica dioica, U. urens.*
O - *Ficus carica, Padus laurocerasus.*

REFERENCES

Andrassy I. *Free-living nematodes of Hungary.* Vol. 3. Budapest, 2009, 608 pp.

Decraemer W. The family Trichodoridae: stubby root and virus vector nematodes In: *Developments in plant pathology,* 6. Dordrecht-Boston-London 1995, 365 pp.

Decraemer W. and Baujard P. Additions and corrections to: De Craemer The family Trichodoridae . *Fundamental and Applied Nematology* 1998, 21, 207-212

Goodey T. *Soil and freshwater nematodes.* 2nd ed., rewriten by J.B. Goodey. London.1963. 544 pp.

Romanenko N.D. *Plant nematodes of family Longidoridae as vectors of plant viruses.* Moscow. "Science", 1993, 284 pp. [in Russian]

Romanenko N.D. *Plant nematodes of family Longidoridae as vector of plant viruses, their relationships with nepoviruses and methods of their control on fruit trees, berry plants and vineyards* (fauna, taxonomy, ecology, harmfulness, virus – vector relationship, plant protection). D.S.C. thesis. 1994. Institute of Helmintology of RAS. Moscow. 1994. 84 pp. [in Russian]

Romanenko N.D. The study of fauna of virus vector nematodes of families Longidoridae and Trichodoridae. In book: "Systematics, taxonomy and fauna of parasites"/ *Materials of science conference.* Moscow 22-24 October 1996. M.1997, P. 103-104 p. [in Russian]

Romanenko N.D. The nematodes as virus vectors. In book: "Applied Nematology". Chapter 6. Moscow. *"Science".* 2006. P. 122-161. [in Russian]

Romanenko N. D. The discovery of complex infections of viruses and nematodes on potatoes under conditions of plant cenoses in the Moscow region. *Abstracts of XXVII International Symposium of European Society of Nematologists* (Rome, 14-18 June, 2004) Rome. 2004. P. 84.

Siddiqi M.R. Phylogenetic relationships of the soil nematode orders Dorylaimida, Mononchidae, Triplonchida and Alaimida with a revised classification of the subclass Enoplia. *Pakistan J. Nematol,* 1983, 1, 79-110.

Taylor C.E., Brown D.J.F. *Nematodes vectors of plant viruses.* Wallingford, UK, CAB International.1997. 286 pp.

Thorne G. *Principles of Nematology.* McGraw Hill Book Co., N.Y., Toronto, London. 1961. 553 pp.

In: Nematodes
Editor: Lee M. Davis

ISBN: 978-1-62948-764-9

Chapter 4

ANALYTICAL METHODS FOR CHEMICAL CHARACTERIZATION OF NEMATODES

Zoltán Sávoly[1,*], István Péter Nagy[2,†] and Gyula Záray[1,‡]
[1]Department of Analytical Chemistry, Institute of Chemistry, Eötvös Loránd University, Budapest, Hungary
[2]Department of Zoology and Animal Ecology, Szent István University, Gödöllő, Hungary

ABSTRACT

Nematodes have a key importance in most aquatic and terrestrial ecosystems, independently of their lifestyle. Moreover, these worms play an important role as bioindicators in the investigation of organic and inorganic chemical pollutants as well. The free-living nematodes are used in biomonitoring due to their specific sensitivity and tolerance patterns towards changes in the environment, for example metal pollution of their habitat. For this reason, broadening our knowledge on their metal uptake and release characteristics will increase our understanding on the various roles they play in contaminated habitats by providing explanations to their stress responses.

* Zoltán Sávoly: Department of Analytical Chemistry, Institute of Chemistry, Eötvös Loránd University, H-1117 Budapest, Pázmány Péter sétány 1/A.
† István Péter Nagy: Department of Zoology and Animal Ecology, Szent István University, H-2100 Gödöllő, Páter Károly utca 1.
‡ Gyula Záray: Department of Analytical Chemistry, Institute of Chemistry, Eötvös Loránd University, H-1117 Budapest, Pázmány Péter sétány 1/A.

Free-living nematodes are microscopic in size and have a relatively short life-cycle. Due to these reasons, and because of their quite complex life processes, they are used as biological model organisms. In order to follow the uptake, distribution, detoxification and biomineralization processes in these animals, special analytical techniques are necessary due to the microscopic size (e. g. Ø=50 μm) of most free-living nematode species.

In this review, the sample preparation procedures for elemental analysis of individual nematodes are demonstrated. Methods which preserve the original structure and enable the investigation of distribution and speciation of micro- and trace elements in small biological samples are discussed. A brief introduction will be provided concerning the microanalytical techniques used for the determination of trace and microelement content of nematodes, such as atomic absorption spectrometry, inductively coupled plasma atomic emission and mass spectrometry, total reflection X-ray fluorescence spectrometry. Furthermore, analytical methods applied for the investigation of elemental distribution (scanning and transmission electron microscopy, synchrotron radiation induced X-ray fluorescence spectrometry) and chemical speciation (X-ray absorption near edge structure, extended X-ray absorption fine structure) are reviewed. In addition to the principle of these instrumental techniques, their analytical figures of merit are also highlighted.

A few characteristic analytical results and observations will be demonstrated e.g. element uptake routes (cuticular or intestinal), metal binding by different organs, biotransformation of heavy metals and nanoparticles, biogenic mineral generation within the body of nematodes, as well as the passage of polluting elements.

The information obtained by these analytical techniques will increase our knowledge on biological effects of polluting heavy metals and may provide explanation on the stress responses of nematodes, these biological indicators and regulator organisms of very high ecological importance.

INTRODUCTION

Nematodes are common participants of environmental investigations. Both the parasitic [1] and free-living [2] species may be used as bioindiactors in the course of the investigation of the pollution of a certain habitat. Although, the elemental analysis is used for investigation of parasitic nematodes [3], often their host is also analyzed and interesting scientific questions can be explained, such as natural bioremediation [1], this chapter focuses on free-living nematodes. The main reasons for this are i.) the important role of free-living nematodes in bioindication and ii.) the difference between the analytical tasks relevant for studying parasitic and free-living nematodes. The latter refers to the fact that most free-living nematodes are microscopic, while parasitic ones are macroscopic. Thus, investigation of free-living nematodes is a typical microanalytical problem, unlike the case of parasitic animals. Nematodes can be used as bioindicators for several reasons [4], the most important are as follows: they are ubiquitous, have direct contact with their environment through their semipermeable cuticle, respond rapidly to disturbances. Microscopic nematodes can be used as biological model organisms due to their small size, however, quite complex life processes. The free-living nematode, *Caenorhabditis elegans* has been used for this purpose since 1963. It played very important role in the description of numerous biological processes [5, 6].

The aim of this chapter is to show what sort of information can be acquired using special microanalytical techniques, and to demonstrate the widespread applicability of these methods. Analytical techniques which enable the determination of micro- and trace element concentrations in microscopic samples may be used to investigate the effects of treating parameters (e.g. concentration of the treating agent, duration of the treatment) on toxic element uptake, as well as the effect of treatments on the concentration of essential elements. So, uptake profile and time-course can be investigated by these techniques. Elemental distribution measurements help to clarify the role of different nematode organs in pollutant uptake and binding, as well as to establish whether biomineralization takes place in a certain case or not. Furthermore, using the most modern analytical methods, information can be obtained concerning chemical environment, in fact speciation without using difficult and time-consuming separation techniques. In order to show the need and applicability of the presented techniques, the most frequently used methods for investigation of free-living nematodes are introduced briefly and references are provided for more information. The techniques discussed herein are as follows: Atomic Absorption Spectrometry (AAS), Inductively Coupled Plasma Atomic Emission (ICP-AES) and Mass Spectrometry (ICP-MS), Total Reflection X-ray fluorescence (TXRF) spectrometry, Scanning and Transmission Electron Microscopy (SEM and TEM) and related techniques (Electron Probe Microanalysis (EPMA) and Focused Ion Beam (FIB)), Synchrotron based techniques such as Synchrotron radiation induced X-ray Fluorescence (SXRF) spectrometry, X-ray Absorption Near Edge Structure (XANES) and Extended X-ray Absorption Fine Structure (EXAFS) methods.

Microanalytical Techniques Applied for Quantitative Elemental Analysis of Biological Samples

Atomic Absorption Spectrometry

Principle and Instrumentation

In the atomic and molecular spectroscopy, absorption is a physical process, in which a chemical species in a transparent medium attenuates electromagnetic radiation at characteristic wavelengths. According to the quantum theory, atoms, ions and molecules have a given set of energy states. At room temperature, most elementary particles are in ground state. When photon passes near an elementary particle, an absorption process takes place if the photon energy matches exactly the energy difference between the ground state and one of the higher energy states of the particle ($M+h\nu \rightarrow M^*$). Due to this energy transfer, the particle is excited. However, the lifetime of the excited particles amounts only to 10^{-8}-10^{-9} s, and the excited species relaxes to its ground states transferring its excess energy to other atoms or molecules in the medium ($M^* \rightarrow M$ + heat). This process results in only a small rise of temperature of the surroundings. Another way for relaxation is the fluorescent or phosphorescent reemission of radiation.

In case of atomic absorption spectrometry, the basic instruments contain primary light sources, atomization unit, monochromator, detector and computer for data evaluation (Figure 1.)

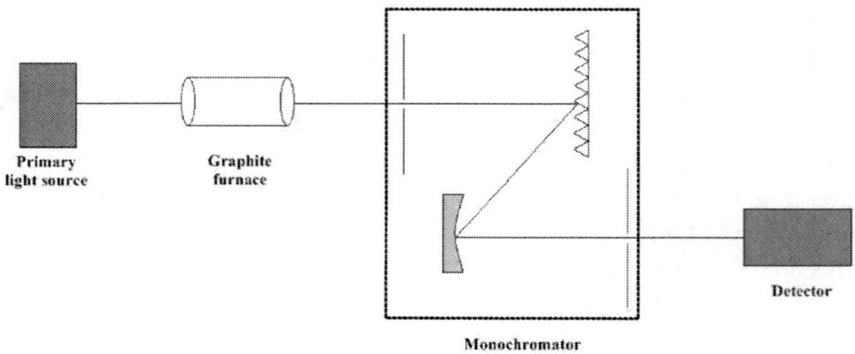

Figure 1. Set-up of the atomic absorption spectrometer.

For primary photon source, mostly hollow cathode lamps (so-called line sources) are used, where the cathode is prepared from selected metals or alloys. These glow discharge tubes emit narrow spectral lines. For these purposes laser diodes can also be applied. In order to produce gaseous medium from the samples to be analyzed, flame and electrothermally heated graphite furnace atomization units were developed. For the microanalytical investigation of low amount biological tissues or even nematodes, the solid sampling graphite furnace technique offers a simple way [7]. The sample on a graphite platform can be introduced into the graphite furnace. The heating program has four steps: drying, ashing, atomization and purification. The drying step at 105 °C is followed by the ashing, namely the removal of the organic matrix. During this step, it is recommended to apply a relatively low heating rate to avoid losses of analytes. However, for atomization, a rapid heating is necessary to achieve a high signal-to-background ratio. In the last step, the graphite furnace is heated up to 2800 – 3000 °C in order to remove all contaminants before the introduction of a new sample. Applying chemical modifiers to the sample, the losses of analytes or interferences can be eliminated. Using the so-called cold vapor (CV) technique, determination of mercury does not need atomizer due to its high vapor pressure at room temperature [8]. To receive accurate analytical results, an efficient background correction is necessary applying deuterium lamp, Zeeman or Smith-Hieftje background correction techniques [9].

The analytical capability of AAS method was drastically improved by the development of the high resolution continuum source atomic absorption spectrometer (HR-CS-AAS) [10]. This analytical system instead of hollow cathode lamps contains as primary light source only a xenon short-arc lamp emitting a continuous spectra in the typical AAS wavelength range (189-700 nm). The high resolution double monochromator with a spectral bandwidth per pixel of 1.6 pm at 200 nm and the charge coupled device (CCD) array detector with 200 pixels guarantee the accurate time-resolved measurement of atomic absorption and simultaneously the background intensity.

To eliminate the risk of contamination and losses of analytes during the sample preparation step, as well as to reduce the time demand of analysis, the direct solid sampling technique was introduced in the analytical practice of GF-AAS [7]. In case of different matrices, it was demonstrated that the solid samples could be analyzed using aqueous standards for calibration. However, depending on the matrix, several spectral interferences that could not be handled by the background correction systems available for the line sources AAS, hampered the accurate analysis.

This problem was solved by the application of HR-CS-GF-AAS technique, owing to its perfect background correction principle.

Analytical Figures of Merit

Since the biological materials usually can be digested easily and rapidly for their elemental analysis, the conventional FAAS [11-13] or recently the HR-CS-AAS with flame atomization [14, 15] are recommended if the sample volume amounts to a few mL.

In case of low-volume samples, instead of the flame atomization, graphite furnace atomization technique should be applied, however, the analytical information provided by this method is restricted only for a few elements [16, 17]. The direct solid sampling combined with the HR-CS-GF-AAS can be successfully applied for determination of cobalt [18] and lead [19] in several biological samples. The detection limits of FAAS technique are in the concentration range of 1-100 ng/mL. Due to the longer residence time of analyte atoms in the atomizer, the GF-AAS method has lower detection limits (0.01-1 ng/mL) than the FAAS. The dynamic range is relatively small, it amounts to two orders of magnitude.

Inductively Coupled Plasma Atomic Emission and Mass Spectrometry

Inductively Coupled Plasma As Photon and Ion Source

The inductively coupled plasma [20] is an electrodeless discharge in a gas (mostly argon) at atmospheric pressure, generated by electromagnetic field with a frequency of 27.12 or 40.7 MHz. A water-cooled coupling coil located around the quartz plasma torch (Figure 2.) is connected to a radio frequency generator. The torch consists of three coaxial tubes with gas by a side tube entering tangentially, creating a vertical flow.

The outer gas flow, termed the coolant flow, protects the tube walls and acts as the main plasma support gas, usually of 10-15 L/min. The small diameter (~1.5 mm) of the injector tube is important to produce a high velocity jet of gas, by which the sample aerosol can be introduced into the plasma forming a cooler axial channel. To initiate the discharge, free electrons are generated by spark from Tesla coil, and the electrical energy applied to the coil is converted into kinetic energy of electrons. The free electron path prior to the collision with an argon or analyte atom, to which its energy is transferred, is only about 1 μm, thus the plasma is heated forming a bright discharge.

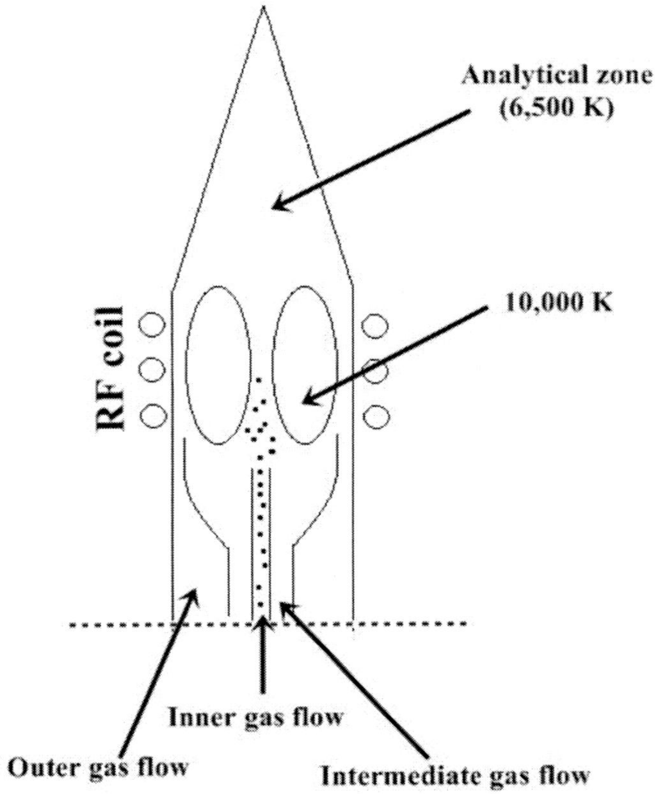

Figure 2. Scheme of the plasma torch.

Due to the skin effect, the plasma has a toroidal form; therefore, the wet or dry aerosol particles can be introduced into the plasma with a good efficiency resulting in a self-absorption free source for the atomic emission spectrometry. The temperature in the induction region amounts to 10 000 K, however, in the central channel above the mouth of the torch, the gas kinetic temperature changes between 5 000 and 7 000 K. This temperature is high enough to atomize the aerosol particles and to excite and ionize the analyte atoms.

For the introduction of liquid samples into the ICP, the solutions are nebulized by pneumatic, ultrasonic or hydraulic high pressure nebulizers (HHPN) combined with spray chambers. The cut-off diameter of spray chambers generally amounts to 5 μm. The nebulization efficiency (volume of solution transported to the ICP/volume of solution transported to the nebulizer) is relatively poor (1-3%) for pneumatic nebulizers (Meinhard, cross flow, glass frit) and better for ultrasonic (~15%) and the highest for the HHPN (~50%).

In the last two cases, the application of a desolvation unit is necessary to eliminate the loading of the plasma with significant amount of solvents. For low-volume (10-50 μl) samples, microconcentric nebulizers [21] or the flow injection method [22] is recommended. In order to eliminate the digestion of solid samples, it is possible to generate dry aerosols by electrothermal vaporization (ETV) of highly volatile elements (e.g. As, Cd, Pb, Zn). Figure 3. demonstrates the set-up of such a unit [23]. The wet biological samples (<100 mg) in a graphite boat may be introduced into the graphite furnace [24]. Following the drying (105 °C) and the ashing steps at 500-600 °C, the temperature of the graphite boat is rapidly increased to 1 600 °C, resulting in hot vapors of volatile elements which are mixed with cool Ar gas forming a well-transportable dry aerosol.

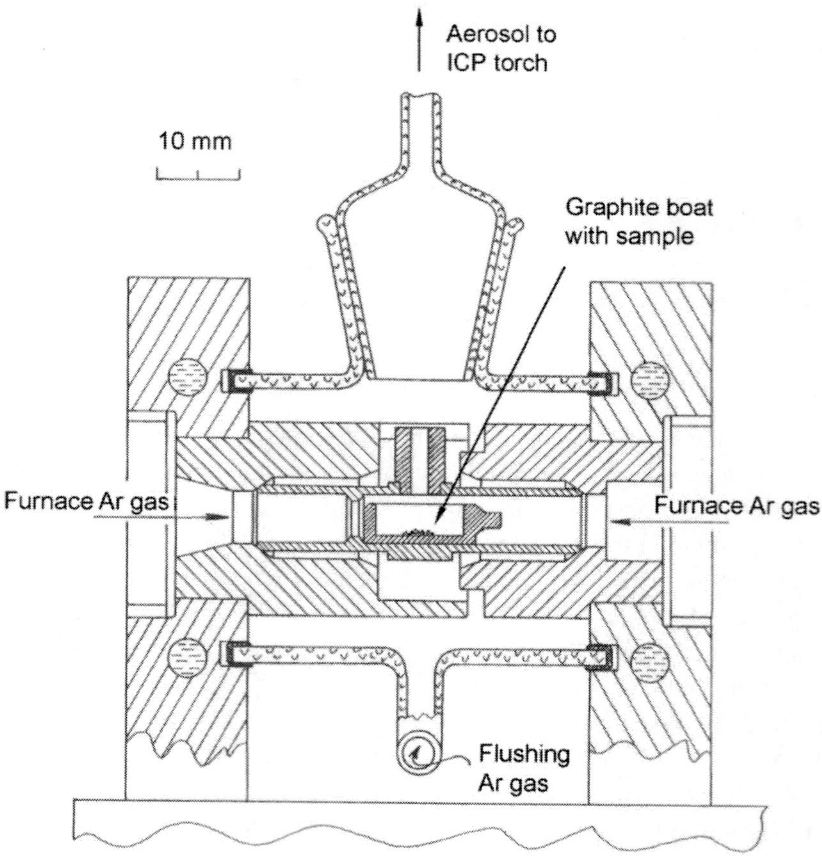

Figure 3. Scheme of the ETV unit.

In order to eliminate the transport losses between the ETV unit and the ICP, it is recommended to apply chemical modifiers, e.g. sodium selenite. In this case, the formation of dry aerosols is supported also by the chemical condensation of vapors [25].

For the excitation and ionization processes in the analytical channel of ICP, first of all collisions of analyte atoms with high-energy electrons or metastable argon atoms are responsible.

The analytical information about the investigated elements can be collected by measuring the emitted photons at the characteristic wavelengths of the elements or by detection of ions of stable isotopes applying quadrupole, time-of-flight or high-resolution mass spectrometers. The ions are sampled from the plasma through a water-cooled interface consisting of a sampler and a skimmer cone with holes of 1.0 and 0.8 mm, respectively. These cones are generally made of nickel, however, for investigation of reactive samples platinum-iridium cones are preferred. Figure 4. demonstrates the schematic structure of the interface unit located between the plasma and the mass spectrometer.

Independently of the detection system (atomic emission or mass spectrometer), there are different spectral interferences resulting in errors of analytical data. In case of ICP-AES, overlapping of spectral lines of different elements is the most critical phenomenon.

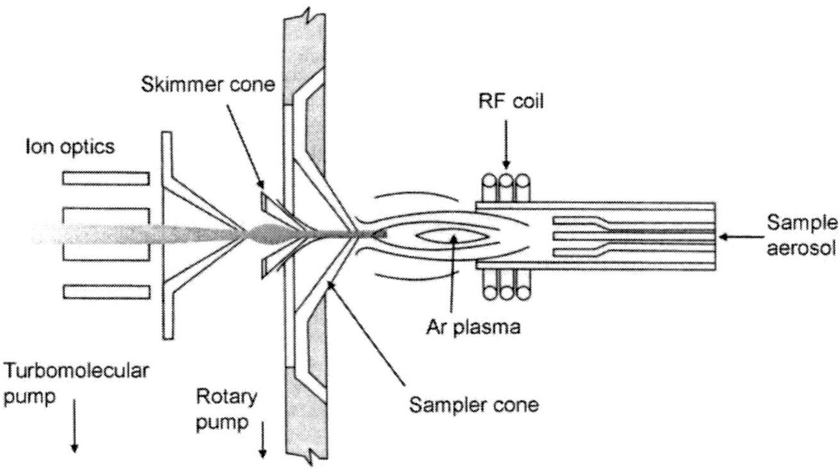

Figure 4. Scheme of the interface located between the ICP torch and the mass spectrometer.

Therefore, it is recommended to apply a monochromator to check the spectral interferences at the wavelengths of the analytical lines in the multichannel atomic emission spectrometers or a multiple line detection for all the investigated elements. Applying mass spectrometers, the ionized stable isotopes of analyte elements are detected at different m/z values. Unfortunately, this technique also suffers from spectral interferences originating from different sources [26]. The first one is the isobaric overlap. This phenomenon occurs if two elements have stable isotopes of essentially the same mass. Fortunately, all elements have minimum one isobaric interference-free isotope. However, their abundance in many cases is not the highest one. The second source of spectral interferences is the occurrence of doubly charged ions. They generate a number of isotopic overlaps at half of the mass of the parent elements. The third group includes the polyatomic ions formed of the atoms of the plasma gas, the solvents (mostly water and acids), the matrix and the analyte elements. In order to eliminate these spectral interferences, there are several possibilities: application of high-resolution mass spectrometers (R=10000); use of collision cells to destroy the polyatomic ions; mathematical correction; modification of plasma parameters (e.g. application of helium instead of argon); removal of disturbing matrix elements prior to ICP-MS measurements.

Analytical Figures of Merit

Using the ICP-AES or ICP-MS techniques, about 70-75 elements can be determined simultaneously and the quantification limits for solutions are in the ppb (ng/mL) or ppt (pg/mL) range. Considering the dilution factors, the quantification limits of inorganic compounds of solid samples are in the ppm (μg/ml) or ppb range. It should be emphasized that the dynamic ranges of ICP-AES and ICP-MS methods amount to 4-5 and 7-8 orders of magnitudes, respectively. This means that the ICP-MS makes it possible to determine major, minor and trace elements simultaneously. Various biological samples can be analyzed both by ICP-AES [27-29] and ICP-MS [30-33].

Total Reflection X-Ray Fluorescence Spectrometry

Principle of X-Ray Fluorescence and Total Reflection

XRF spectrometry is a well-established method for elemental analysis. The determination of chemical composition through the measurement of XRF spectra has become a routine analytical method, employed in a wide variety of research areas ranging from material research to biomedical and environmental sciences. XRF is a non-destructive method, for which only simple sample preparation is necessary. The elements which can be analyzed by commercial laboratory scale equipment range between Z=14 (Si) and Z=92 (U). The analysis of elements below Si requires specific conditions: detector with thin entrance window, measurements done in vacuum, special sources for excitation (e.g. synchrotron radiation or special X-ray tubes with low Z anodes) [34]. The basic phenomenon of XRF spectrometry has two steps: the first is the ionization process in one of the inner shells caused by high energy X-ray photons (generated by X-ray tubes, radioactive isotopes or synchrotron), and the second is the neutralization of the ionized shell by an electron jumped down from an outer atomic shell having lower binding energy. The difference of the two binding energy values will be carried by a fluorescence photon that is emitted during the relaxation process. Due to the quantified binding energies of the different electron shells in the atoms, the emitted fluorescence photons have a well-defined energy. The electron transitions are regulated by special selection rules. The typical transitions are demonstrated in Figure 5. These characteristic lines in the X-ray spectra (Figure 6.) provide both qualitative and quantitative information on the elements contained in the analyzed material [35]. For detection of fluorescence photons, both wavelength and energy dispersive systems are available. In the conventional XRF equipment, the primary X-ray beam arrives to sample surface under angle of 45° and the penetration depth changes in the μm-range depending on the physicochemical properties of the sample and the energy of the primary photons. Due to the relatively long pathway of the primary and the fluorescence photons in the sample, the matrix effect has a strong influence on the analytical signals. This disturbing phenomenon was eliminated by Yoneda and Horiuchi [36] in 1970 by developing the total reflection X-ray fluorescence (TXRF) method. According to the Snellius-Descartes law, in case of a significantly low value of angle of incidence (<0.1°, below the critical angle), the exciting X-ray is totally reflected [37] from the air/sample holder interface. The total reflection of the exciting radiation improves greatly the analytical parameters of the XRF spectrometry. Figure 7. shows the set-up of the TXRF spectrometer.

The advantages of TXRF against the conventional geometry are as follows: 1.) both the incident and the reflected radiation excite the samples; 2.) there is a lower background intensity due to the reduced scattering of the exciting radiation; 3.) the detector can be positioned closer to the sample surface and hereby the efficiency of fluorescent photon collection is much higher.

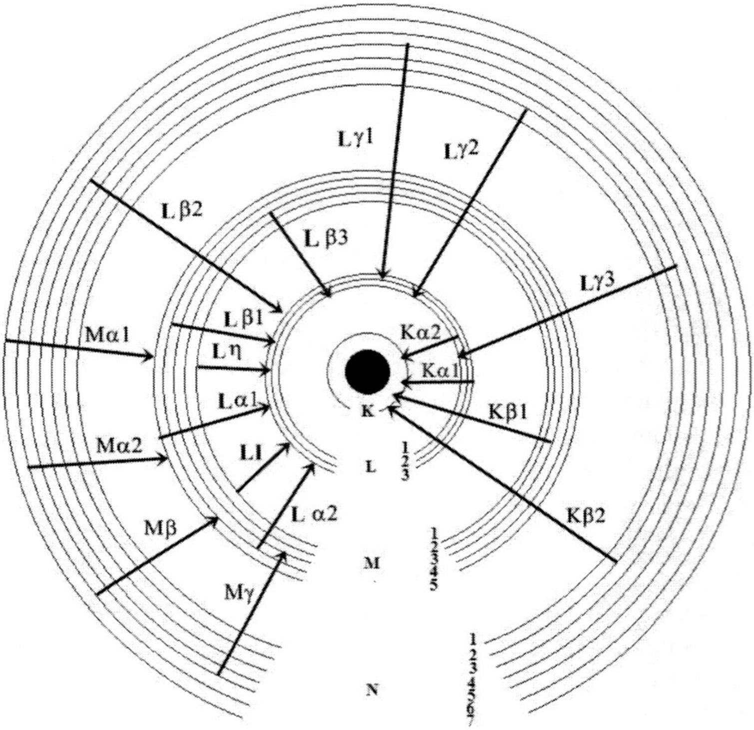

Figure 5. The most important allowed X-ray transitions. The black circle stands for the nucleus, while K, L, M and N symbolize the electron shells.

Performance and Analytical Characterization

TXRF spectrometry needs sample holders possessing high reflectivity. They can be made of several materials, with the most common ones being: quartz glass and plexiglas.

The former is considerably expensive, but very resistant, therefore enables the analysis of reactive materials (i. e.: acids, organic solvents) and can be cleaned and reused, while Plexiglas is inexpensive, but can be used just once, and enables only the analysis of water solutions and suspensions. The application of quartz carrier plate, however, hampers the determination of silicon.

Using TXRF spectrometry, several microliters of liquid or a few micrograms of solid samples may be analyzed. The sample should be deposited on the sample holder and dried. The key of the successful TXRF analysis is the sample thickness. The recommended values for different matrices are in the range of 10 nm-12 μm.

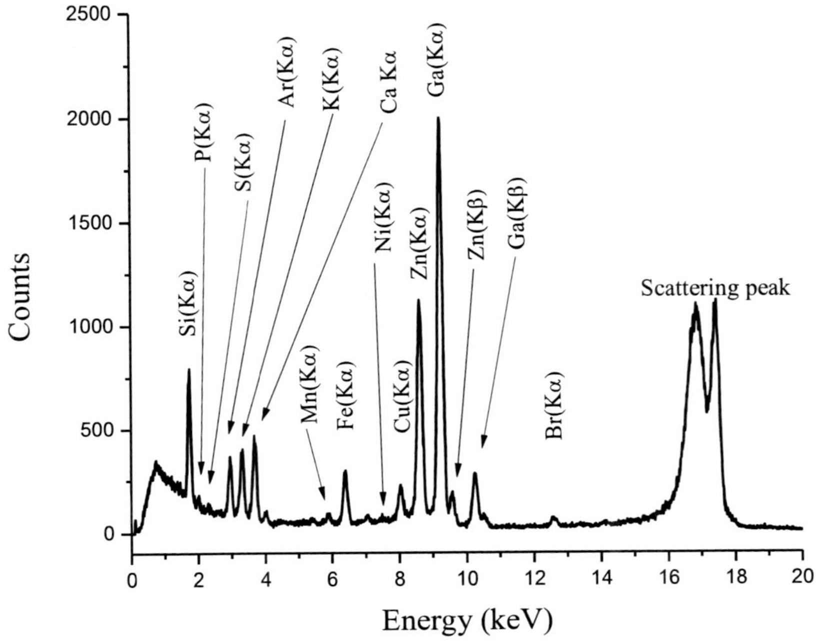

Figure 6. X-ray fluorescence spectrum of a *Xiphinema vuittenezi* nematode.

The lowest thickness is allowed for metal layers and the highest for organic materials [34]. Despite this limitation, TXRF can be widely used for the analysis of several types of samples.

In most cases, the sample is digested and analyzed as a liquid, however, TXRF enables also the investigation of suspensions, atmospheric aerosol samples collected directly on the sample holder [38], individual particles, pigments from oil paintings (collected by cotton buds) [39], sections of biological samples prepared by microtome [40].

Sample digestion can be performed directly on the sample holder using small amount of concentrated acids; another possibility for matrix removal is ashing, using cold oxygen plasma [41].

Internal standardization can be used as a simple calibration method for multielement analysis. An internal standard is added to the sample in a certain concentration and homogenized with it, then the analytical signal is measured for the standard and the analytes.

Knowing the relative sensitivity of the investigated elements related to the standard, the analyte concentrations can be calculated.

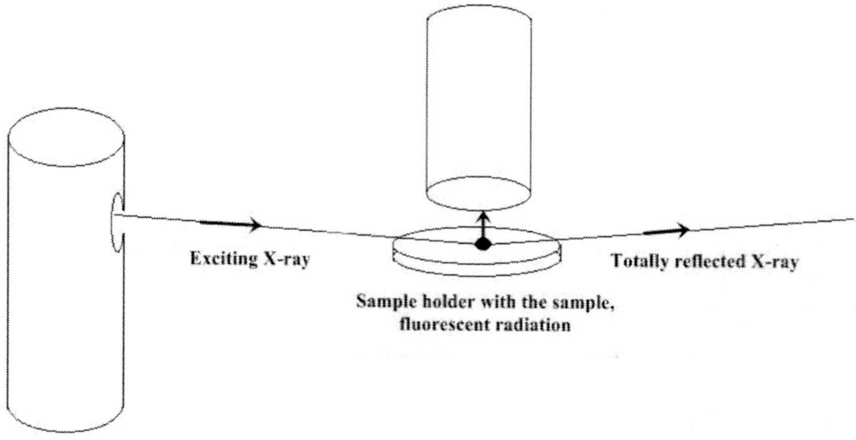

Figure 7. Set-up of the TXRF spectrometer.

The relative sensitivity is determined by preliminary experiments, measuring the intensities for a sample containing both elements in given concentration. The relative sensitivity depends only on the applied X-ray tube, the voltage, and the filter or monochromator used for modifying the exciting radiation; its redetermination is necessary only in case of changes in the above-mentioned parameters.

By means of TXRF spectrometer working in normal air atmosphere, principally more than 75 elements can be determined. The investigation of low-Z elements ($Z<14$) is hampered by the intense absorption of their low-energy fluorescent radiation in air and the higher spectral background in the low-energy range; however, significant developments were achieved over the past years concerning their detection [42]. The dynamic range of TXRF is about 4 orders of magnitude, which can be increased to 6 under special conditions. The absolute detection limits are 1-10 pg for almost all the elements. It means that elements in the μg/L concentration range can be measured even in case of small volume samples (5-10μL).

TXRF spectrometry is widely used for the investigation of biological samples; however, here it is discussed only shortly, for details see [43]. The main advantage of the technique is its small sample-demand. Macromolecules separated from their natural matrix were studied by TXRF, i. e.: proteins [44, 45], nucleic acids [46].

Different biological fluids were measured by this technique: urine [47], ([48]); whole blood [49]; serum [50], ([51]); amniotic fluid [52], ([53]); cerebrospinal fluid ([54]); tea [55]; xylem sap of plants [56]. These fluids were analyzed with or without digestion (direct analysis), and the samples studied in the latter way are marked in parentheses. Cells were also investigated by a conventional method, depositing the digested samples on the sample holder [57], or these samples were digested directly on the sample carrier plate [58]. For investigation of animal [59], plant [60] and human (mainly cancerous [61]) tissues, TXRF spectrometry has been widely applied.

Methods Used for the Determination of Elemental Distribution and Speciation in Biological Samples

The modern instrumental methods enable not only the determination of concentration of micro- and trace elements, but their distribution and speciation can be measured within the sample. For elemental distribution measurements, the non-destructive X-ray spectrometric techniques (EPMA, SXRF) are often applied. In case of such investigations, more serious preparations are needed for biological samples comparing with other ones, for example samples of materials or earth sciences. It is due to the fact that biological samples are more sensitive to the circumstances given during the investigations (i. e.: high vacuum, bombardment with high intensity radiation, such as electron beam or X-ray radiation). If the structure of the sample or the speciation of the investigated elements is significantly altered during the measurements, the results are not related to the original state of the investigated biological system. Many techniques were developed earlier for the investigation of the structure of biological samples with SEM. Most of these techniques are allowed to be used for analytical investigations as well. There are techniques which need a special electron microscope. In this case, EPMA measurements may be carried out, while in case of other type of techniques, sample preparation is carried out prior to the measurements, which enables them to be used for both EPMA and synchrotron measurements. The main reason of the increased sensitivity of biological samples is their relatively high water-content. It is incompatible with high vacuum or high intensity radiation, in most cases hydrated samples suffer serious deterioration during measurements.

The removal of water content is necessary, especially in case of investigations, which need preservation of the original structure such as SEM, EPMA and SXRF methods [62]. Special modes of modern SEM devices enable investigation of biological samples.

The environmental SEM works using water steam environment in the pressure range of 130-2700 Pa. The sample is measured at the temperature close to the triple point of water, hereby 100 % relative humidity can be achieved [63].

A further special mode of SEM is the low-temperature SEM, in which case the sample is refrigerated (cryofixed) and measured [64]. Water removal of the sample by retaining its original structure can be achieved using the techniques as follows: supercritical drying (water content of the sample is substituted with carbon-dioxide and that is removed at critical temperature) [65]; quick-freezing followed by lyophilization (freeze-drying in vacuum) [66]; chemical fixation (using several chemicals, for example glutaraldehyde or osmium tetroxide) [67].

Electron Microscopy-Based Techniques

The Electron-Matter Interaction

The operation of electron microscopes is based on the electron-matter interaction (Figure 8.). In the SEM [68] and TEM [69] equipment, the sample is bombarded by high-energy (5-30 keV) electrons generated by an electron gun, containing a tungsten filament or a lanthanum hexaboride cathode. The appropriate positioning (focusing) of the electron beam is made by special electromagnetic lenses, the so-called electron optics, these names refer the similarity with optical microscopy.

The difference between SEM and TEM is mainly technical. In case of SEM, the picture is taken pixel by pixel, while for TEM, all the pixels are irradiated simultaneously. The morphological information for SEM originates from the sample surface layer, while the TEM needs considerably thin samples (<100 nm), through which electrons can be transmitted.

Therefore, the latter technique requires special sample holders and more complicated sample preparation.

In case of SEM, the particles generated due to electron-matter interaction are detected on the side of the primary radiation (related to the sample), while TEM detect the transmitted or the scattered electrons on the other side. Both the secondary and backscattered electrons are used for picture generation.

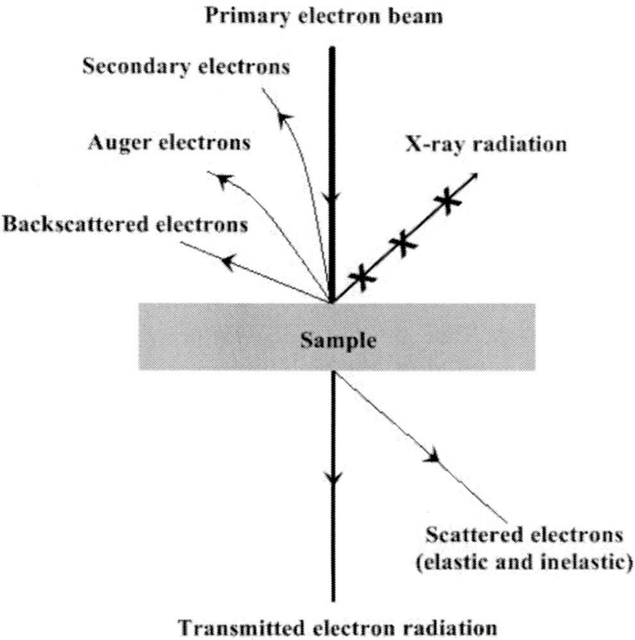

Figure 8. The electron-matter interaction.

Secondary electrons are emitted from the sample due to the electron-matter interaction, while the backscattered ones are high-angle scattered particles of the primary beam. The backscattered electron detection provides picture with Z-contrast, because with increasing Z, the back-scattered electron intensity increases. SEM needs conductive samples, therefore insulating ones should be covered with a thin layer of conducting material (gold or carbon).

The other opportunity is measuring at low vacuum environment (0.01-130 Pa) or using the above-mentioned environmental SEM. The fluorescent X-ray radiation generated by the high-energy electrons can be used for analytical purposes (EPMA method, see the next subchapter). Auger electron generation is alternative to X-ray radiation for energy emission after electron cloud rearrangement. In this case, instead of electromagnetic radiation, electrons are emitted from the excited sample surface [70]. Electron energy loss spectrometry (EELS) can be used in case of TEM, for analytical purposes. By investigating the loss of the energy of the incident beam, information can be gained on the elemental composition of the sample, due to the ionization caused by the beam [71].

The modern SEM instruments enable performing scanning transmission electron microscopy (STEM); it is transmission spectrometry in terms of detection, but picture is generated pixel by pixel.

Electron Probe Microanalysis

For EPMA, only an X-ray detector is needed inside the electron microscope. The modern electron microscopes are equipped with energy dispersive detector. During EPMA investigations, the sample is irradiated by electrons and the released X-ray radiation is detected. The rapid qualitative or semi-quantitative analysis of quite small sample volumes (μm^3) can be realized for Z>3 elements. Due to the high vacuum applied inside the electron microscope (10^{-2}-10^{-4} Pa), the low-energy fluorescent photons emitted by low-Z elements can also be detected. The upper limit concerning the energy of detectable X-ray lines is limited by the energy of the exciting electron beam. The quantitative analysis needs standard samples with the same matrix, what the sample to be analyzed has. Since the matrix of the sample and the standard differs in most cases, several mathematical methods were developed to eliminate the matrix effect. The most known of those is the so-called ZAF correction [72]. "Z" denotes the atomic number dependence, where two factors should be considered: the backscattering efficiency and the electron-stopping power. The former one increases, while the latter one decreases with increasing Z. "A" considers the absorption of the characteristic radiation. "F" denotes the fluorescence, the characteristic radiation can suffer X-ray fluorescence increasing the signal of the excited lower-Z element. Obviously, it is significant in case of elements having lower atomic number. The method is iterative – since Z, A and F values depend on the matrix and its constitution is not known – there is a first approximation and it is being improved.

Beyond the application for traditional analytical measurements, EPMA can also be used for study of elemental distribution. Although the excitation is carried out by a narrow electron beam (Ø>1 nm), the lateral resolution of the elemental map is significantly higher due to the larger excited volume. The resolution of the method is 0.1-1 μm and the detection limits are in the range of 100-1000 μg/g. Therefore, this method is not suitable for trace analysis. The analytical depth (from where information can be still gained) amounts to 0.1-1 μm. The presented data concern biological samples [73].

Focused Ion Beam Technique

The modern SEM devices contain so-called FIB unit, which broadens the applicability of the instrument.

FIB is consisted of Ga^+ ions in most cases, which is prepared by the so-called liquid metal ion source [74]. Gallium is stored in a container, it is dosed applying a wolfram needle and electric field, this time ionization takes place as well. The beam is focused by ion optics. The ion-matter and electron-matter interaction are similar. Ions can remove electrons, thus electrons and X-ray photons are generated, in addition secondary ions and ablated particles are generated as well.

Consequently, FIB technique has widespread applications. By detecting the emitted electrons, pictures can be taken similarly to the SEM technique [75], but in this case secondary ions can be detected as well. The ablation of the sample allows for the preparation of considerably smooth sections [76]. By changing the ion current and the mode of scanning, several processes can be carried out on solid samples: cutting (with higher ion current), cleaning (with lower currents).

TEM samples can be prepared by FIB as well [77]. Another application is the chemical vapor deposition [78]. For this purpose a precursor gas should be introduced into the sample chamber, and due to the interaction of the precursor gas with the ion beam, a pre-defined layer can be deposited. Before application of FIB cut process, it is recommended to form such covering layer on the sample surface, where the part of the sample which is not intended to be cut is prevented from significant ablation.

Biological Applications

The SEM method has been applied for the investigations of structure of biological samples for a relatively long time [79], however, FIB-SEM system offers more precise information [80].

EPMA is widely used for studying biological samples following an appropriate sample preparation step. A special advantage of EPMA is that it makes possible the simultaneous determination of low-Z elements (C, N, O, P, S, Na, K, Ca) which plays a dominant role in the formation and the structure of biological materials.

In addition to the chemical composition, the 2 or 3D [81] distribution of elements can also be determined. It should be mentioned, however, that the detection limits are relatively high, therefore this technique is not suitable to study the distribution of trace elements. Table 1. summarizes the recent EPMA studies focused on biological samples.

Table 1. Biological samples studied by EPMA

Sample	Method	Sample preparation	References
Alyssum lesbiacum (plant)	SEM; elemental map: Ni, Ca; concentration: Ni, K, Ca, Mg	Freezing, five-day-long drying in vacuum	[82]
Nerium oleander (plant)	SEM; composition at more points: 14 elements	Freezing in liquid nitrogen	[83]
Cannabis sativa (plant)	SEM and TEM, investigation of copper at more points	SEM: supercritical drying TEM: chemical fixation	[84]
Clostridium tyro-butyricum (bacterium)	SEM; elemental map: Ca; detected: P, S	Dehydration with ethanol and supercritical drying	[85]
Chelonia mydas (turtle) eggshell	SEM; elemental map: Ca; concentration: 11 elements	Direct investigation	[86]
*Pinctada margar-itifer*a (pearl-oyster) shell	SEM; elemental map: Na, Cl, Mg, Sr	Polishing, acidic and enzymatic digestion	[87]
Earthworms	SEM; elemental map: 11 elements	Cryofixation	[88]
Earthworms	TEM; concentration: P, S, Cl, K, Ca, Fe, Zn	drying by ice-cold sucrose solution	[89]
Human brain tissue	SEM; investigation of C, O, P, S, Ca, Mg	Fixation with formaldehyde	[90]
Human red blood cell	TEM; Na, K, Fe elemental map and concentration	Freezing, lyophilization	[91]

Synchrotron Radiation-Based Methods

Synchrotron Radiation

Synchrotron radiation is generated, when an ultrarelativistic charged particle is accelerated radially [92].

The first-generation synchrotrons were built to perform particle physics-related experiments, synchrotron radiation-based measurements were carried out as parasitic operation. The second-generation were constructed specifically for using synchrotron radiation. The construction of a synchrotron light source is presented in Figure 9.

The electrons emitted from an electron source are accelerated in a linear accelerator and a synchrotron ring, then they are transported into the storage ring. The electrons are kept on their orbit by magnets, and their energy loss is compensated continuously by using RF field.

Synchrotron radiation is emitted tangentially, nowadays special magnetic devices (third-generation) are used for the optimization of the radiation. Synchrotron radiation covers a wide energy range from the infrared radiation to the hard X-ray.

Its intensity is 13 orders of magnitude higher, than the X-ray radiation generated by conventional X-ray tubes. It has high polarization rate in the plane of the electron orbit, which is beneficial in case of X-ray fluorescence measurements. This radiation in vertical direction is highly collimated, and its pulse-nature makes possible the time-resolved measurements. The energy of the radiation can be easily tuned using a monochromator.

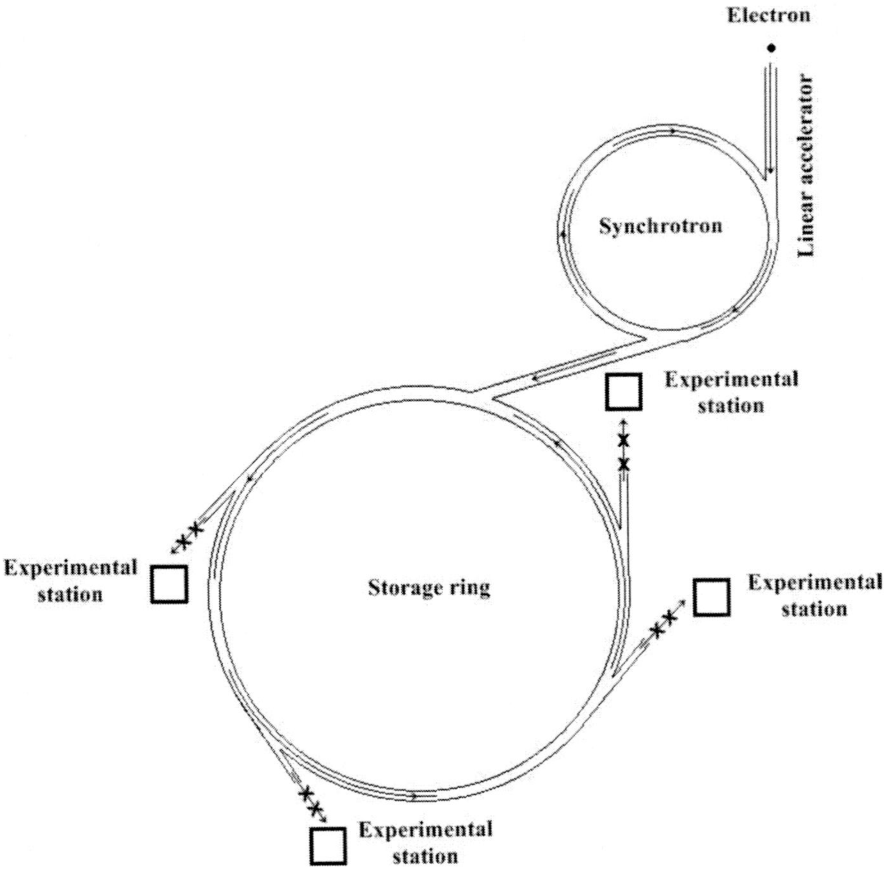

Figure 9. Scheme of the synchrotron light source.

Accordingly, synchrotron radiation has numerous applications, i. e.: tomography [93]; time-resolved X-ray diffraction [94]; inelastic X-ray scattering [95]; X-ray fluorescent spectrometry; X-ray absorption spectrometry. The two latter fields are discussed in the following subchapter.

Synchrotron Radiation Induced X-Ray Fluorescence Spectrometry

Elemental maps can be created, by the focusing of the exciting radiation to a narrow beam (0.1-10 μm). Obviously, the applied method should have high detection power, thus considerably low detection limits for the investigated elements, since the small irradiated area contains necessarily low amounts of analytes, especially in case of trace elements. The above-mentioned properties of the synchrotron radiation allow for the achievement of absolute detection limits in the femtogram range.

Due to the high performance electron optics, elemental mapping were carried out in two [96], and in three dimensions as well [97]. Submicron spot size was produced using one of the optical devices as follows: Kirkpatrick-Baez mirror system [98], compound refractive lenses [99], Fresnel zone plate [100]. The principle of these systems: total reflection, refraction, diffraction, respectively. In case of X-ray fluorescence tomography, the sample is rotated in more steps from 0 to 180° and 2D elemental maps (projections) are detected [101]. The 3D distribution of elements is calculated by using a special algorithm.

Comparing with other techniques, SXRF method has a relatively high analytical depth (100-1 000 μm). Its spatial resolution is 0.1-1 μm, therefore, it is suitable for the elemental analysis at subcellular level [73].

X-Ray Absorption Spectrometry

X-ray absorption spectrum (rate of absorption vs. the energy of the applied X-ray radiation) of a certain element shows the decrease of the absorbance increasing the energy of the exciting radiation, however, the spectrum shows breakpoints, where a high increase in the absorbance is observable. These are the so-called absorption edges (Figure 10. A). At these edge positions, the energy of the absorbed radiation is equal to the energy needed for the removal of an electron at a certain energy level.

Accordingly, K, L, M edges exist, and the letters denote the shell from which the electron is removed. The position of the absorption edge and the fine structure of the X-ray absorption spectrum (XAS) are characteristic of the oxidation state and the chemical environment of a selected element, respectively.

Due to the high tunability of synchrotron radiation, this energy range can be studied with high resolution, and for the practice two parts of X-ray absorption spectrum are distinguished (Figure 10. B).

The fine structure is affected by the neighboring atoms, due to the interference of photoelectron waves. The photoelectrons are removed from the analyte atoms by X-ray radiation and an interference takes place between the exiting photoelectrons and the ones which are reflected from the neighboring atoms. The EXAFS range begins at about 50 eV far from the edge, but it is not a strict rule.

In this region, photoelectrons have relatively high kinetic energy resulting in mainly single scattering processes. Consequently, direct neighboring atoms of the analyte atom, thus the first coordination sphere can be studied. Quantitative investigations are enabled and the absorption process in this energy range can be characterized by an equation [102]. Bond length and angles can be determined as well as the quality of the neighboring atoms.

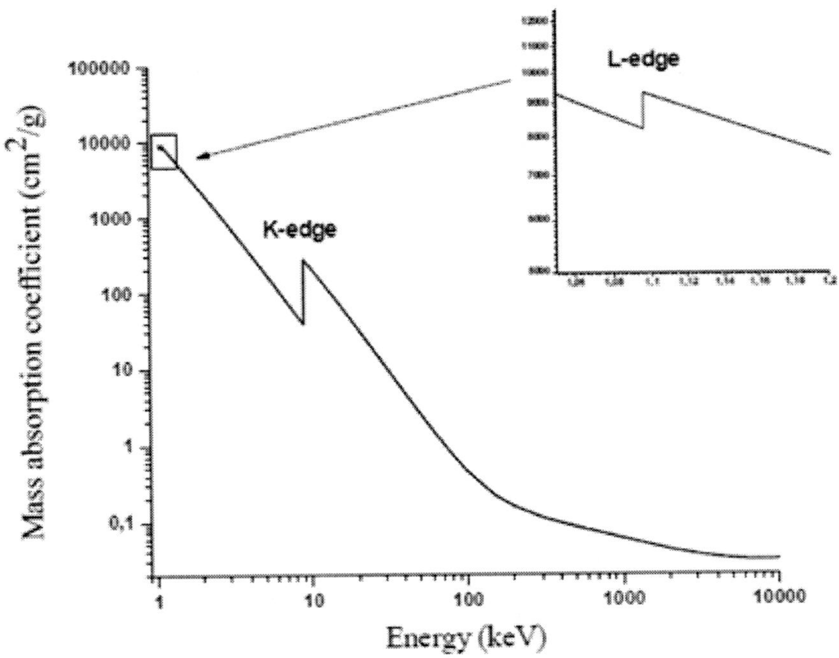

(A)

Figure 10. (Continued)

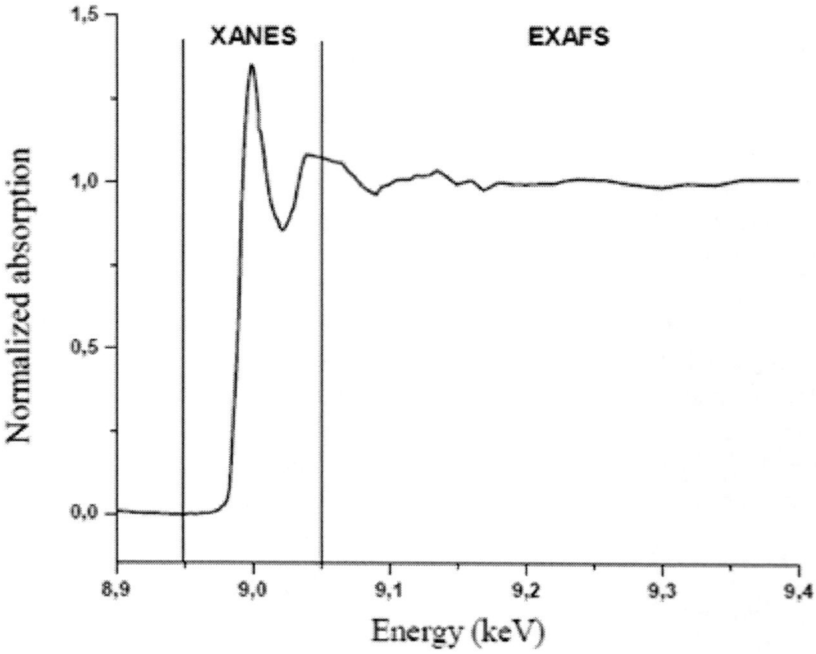

(B)

Figure 10. Low- (A) and normalized high-resolution (B) XAS spectrum of copper and $Cu(CH_3COO)_2$, respectively.

The method has limitations in terms of the latter parameter, because the atomic number dependence is not so sensitive, therefore just those atoms can be distinguished, which have higher Z difference (i. e.: oxygen and sulfur, but oxygen and nitrogen not).

In the XANES range, due to the lower exciting X-ray energy, the photoelectrons have lower kinetic energy, hereby their multiple scattering takes place. It makes the theoretical description of the XANES spectrum a challenge, therefore the XANES technique is mainly used for comparative studies, but continuously improved theoretical models exist [103]. Nevertheless, XANES method is much more sensitive to the chemical environment of the analyte atom, than the EXAFS. In case of many similar compounds, no difference is found in the EXAFS spectrum, but their XANES spectrum differs. The XANES spectrum is in fact a fingerprint of the investigated compound, which enables comparative studies. One of the most important properties of the spectrum is the position of the edge, since it shows significant dependence on the oxidation number of the analyte atom.

The higher is its oxidation state the higher is the energy of the edge, since the higher is the positive charge of the atom, the higher energy is needed for the removal of an electron. Further properties (presence and size of pre-edge peak or shoulder) characterize the geometry around the analyte atom [104, 105]. If the appropriate reference compounds are available, actually speciation investigations can be carried out. In this case, XANES spectrum of the sample is fitted by the linear combination of reference compound spectra [106]. The XANES method has lower detection limits, than the EXAFS.

XAS measurements can be carried out using the conventional absorption geometry (as in case of AAS), however, fluorescence mode can be applied as well. In this case, the exciting energy is being increased during the measurements the same way, however, not the rate of absorption, but the intensity of fluorescent radiation proportional to the absorption rate of the exciting radiation, is measured [107].

Biological Applications

Both the X-ray absorption and the fluorescence techniques have high number of considerable biological application possibilities, therefore only a few examples are presented here.

Goldberg et al. investigated elemental distribution in tree rings, and found that the investigated elements can be classified into two groups. The trend of the elements of the first category (Br, Zn, Cl) was found to correlate with atmospheric precipitation. Consequently, investigating the distribution of one of these elements might be used to reconstruct past climate humidity [108]. Mihucz et al. investigated the distribution of several elements (i. e.: Zn, Mn, Ti) in rice and discovered a 80-μm-wide element rich surface region [109]. Various animal and human samples were investigated, for instance: snail shell [96], tooth [110], several tissues [111, 112]. The distribution of different elements within the same sample can be compared quantitatively with correlation analysis. Furthermore, elemental correlation maps can also be generated using two or three maps having one of the colors as follows: red, green, blue. Summing these pictures, the outcome is a correlation map. At points, where the blend of two or three colors take place, the corresponding elements coexist [113]. The SXRF method allows for the determination of the elemental distribution in 3D [97] as well as at subcellular level[114], and combining these techniques, the 3D investigation of cells becomes possible [115]. The widespread applicability of XAS methods is proven by the high number of types of biological samples as well as the studied chemical elements.

Investigations at the cell level can be carried out not only with SXRF but also with XAS [116]. Recently, the investigation of metal tolerant and hyperaccumulator plants has been a popular topic due to their potential applicability in bioremediation. XANES and EXAFS methods provide an excellent tool for the study of the chemical environment of the accumulated element [117-119]. Another examples for investigated samples and elements: mercury (L_{III}-edge) and sulfur (K-edge) XANES in intestinal bacterium [120]; zinc (K- edge) EXAFS in several tissues of nano ZnO-treated rats [121]; iron speciation (K-edge XANES) in human cancer cells [122]. Strub et al. described a method which is suitable to determine the Cr(III)/Cr(VI) ratio by the determination of the height of the intense pre-edge peak located before the K-edge of Cr(VI) [123]. SXRF and XANES are often used altogether [124] in order to investigate the speciation of the analyte elements at different points of the sample, moreover XANES mapping (investigation of the distribution of a certain species) may be carried out [125].

Microchemical Investigation of Nematodes

Study of Element Uptake

The study of element uptake needs multielemental analytical methods having low detection and quantification limits. For this purpose, the ICP-AES, ICP-MS and the TXRF spectrometry are the most powerful methods. The selection of the analytical method to be applied depends also on the sample demand of the analysis. Analytical techniques working with continuous sample introduction e.g. FAAS, ICP-AES or ICP-MS need high volume solution (3-5 ml) prepared by joint digestion of high number of nematodes. The sample demand of ICP-AES and ICP-MS can be reduced to 10-20 μl applying electrothermally heated vaporization unit. The GF-AAS method needs also only 10-20 μl solution for one analytical step, but its application results in only monoelemental information. For the multielemental analysis of individual worms, the ETV-ICP-AES, ETV-ICP-MS and the TXRF methods are suitable. In case of these 3 techniques, direct solid sampling technique can be applied reducing the risk of contamination.

Helmcke et al. studied the effect of methylmercury on *C. elegans* [126]. High amount of animals (approximately 10,000 for L1 and 900 for L4) was treated with methylmercury, and after washing twice in deionized water, the protein content was determined, in order to express the mercury content as ng

Hg/mg protein. Sample digestion was carried out in a block heater, following the addition of 65% HNO_3. Digested samples were diluted with ultrapure water. Mercury content was determined by high resolution (HR) ICP-MS technique, without using internal standard, applying low resolution (m/Δm=300).

Besides mercury content, other parameters were also investigated (i. e. toxicity, effect on lifespan, swimming rate). It was found that at relatively high internal mercury concentration (max 3 ng Hg/mg protein, that is highly toxic to mammals), only limited effects were observable on the above-mentioned parameters. For example, neurons of the worm were less sensitive to methylmercury than of mammals. Consequently, *C. elegans* provides a good tool for the investigation of resistance to methylmercury. This group had another publication in the same topic [127], studying the role of glutathione (GSH), metallothioneins (MTs) and heat shock proteins (HSPs) which reduced the toxicity of methylmercury. Caito et al. had similar investigations on the same nematode species with methylmercury [128]. Samples containing $4*10^5$ nematodes washed twice in ultrapure water were digested at 100 °C with a mixture of concentrated HNO_3 and HCl, then diluted with ultrapure water, and gold was added as the internal standard. HR-ICP-MS measurements were carried out for determining mercury concentration at medium resolution (m/Δm=4400). The concentration was expressed as ng MeHg/mg worms. The effect of amino acid transporters (AATs) on the MeHg toxicity was studied. Knockdown of AATs encoding genes resulted in reduced toxicity as well as lower internal Hg content. Toxicity was blocked in case of pre-treatment of worms with L-methionine, which suggests the role of AATs in MeHg transport as well.

Lin et al. investigated the positive effect of manganous ion supplementation on several parameters such as speed of development and life span [129]. ICP-MS was used to follow the uptake process of manganese. *C. elegans* specimens were treated with $MnSO_4$ solution. Samples contained exactly 40 worms. Following the dehydration and cooling, the worms were digested in heating block at 98 °C for 2 hours, applying 1 ml of 20% HNO_3. Results were presented as pmoles Mn per worm. Manganese was found to be taken up by the worms, however, it should be noted, that the untreated worms contained about 1 pmol Mn (each). The effects of selenium treatment on the same nematode species were studied by Li et al. [130]. Selenium is known to be an essential microelement, which has a very narrow exposure window between its beneficial and toxic effects. Nematodes were treated with 1, 5 and 20 μmol/L Na_2SeO_3 solutions.

Following the treatments, the nematodes were washed with K-medium five times applying an incubation for 30 minutes between each washing step. Incubation was applied in order the remove excess selenium and bacteria used for feeding from the gut of the nematodes. Sonication was applied to shatter and homogenize the nematodes and following a centrifugation, the selenium concentration of the supernatant was measured by ICP-AES method. Protein content was also determined, and the selenium concentration was presented as selenium/total proteins (μg/mg).The selenium uptake was found to be concentration-dependent. Wan et al. investigated the effect of β-amyloid peptide which is related to the pathogenesis of Alzheimer disease (AD) on the iron content of human cells as well as *C. elegans* [131]. Using ICP-MS method, they established that the transgenic AD-associated nematodes had significantly higher iron content than the wild-type nematodes.

Todays, the engineered nanoparticles play an increasing role in different industrial areas as well as medical applications due to their beneficial properties. During the last decades, two research areas were born: bionanotechnology and nanotoxicology. Nematodes are used as biological model organisms in these fields. Kim et al. investigated the uptake and positive effects of cell-penetrating peptide conjugated platinum nanoparticles [132]. Following a two-day-long treatment, nematodes were washed with M9 buffer three times and the washing solution was removed by centrifugation. One-hundred nematodes were resuspended in the buffer and dissolved by a mixture of HCl, $HClO_4$ and HNO_3 (10, 15 and 15% in terms of final concentration, respectively) at 200 °C. After cooling, samples were resuspended in 5 ml of 1% HNO_3. ICP-MS was applied for the quantification of platinum. Concentration was expressed as platinum content in nanomoles of 100 worms. A dose-dependent platinum internalization was found. The applied nanoparticles were found to be effective antioxidants, hereby induced the extension of lifespan of *C. elegans*. Zhao et al. compared the effect of acute and prolonged uptake of engineered TiO_2 nanoparticles in *C. elegans* [133]. The internal titanium concentration was measured by ICP-MS and expressed related to protein content. Further details are not presented concerning the analytical measurement in this paper. An acute exposure at higher concentration (100 mg/L) and a prolonged one at lower TiO_2 concentration (100 μg/L) were applied. The uptake of titanium and its effect on the intestinal barrier and neurons controlling defecation were investigated directly after the exposure and 48 hours later. In case of acute treatment, the biological changes in terms of the above-mentioned investigations returned to the normal level, while for prolonged treatment no recovery was found.

It is in accordance with the result of the ICP-MS measurements, the titanium content decreased significantly in 48 hours for acute treatment, whereas for prolonged treatment the decrease was only moderate. Yeo et al. investigated artificial paddy microcosm treated with TiO_2 nanoparticles and nanotubes [134]. Titanium concentration of several organisms, sediment and water samples were determined by ICP-MS method. Among others, plant-feeding nematodes (*Meloidogyne* sp.) were also studied. Microwave assisted digestion was used applying nitric acid and hydrogen peroxide solutions. Analytical results in μg/g, furthermore the bioconcentration and bioaccumulation factors were presented. It was established that a considerable transfer of nanomaterials took place from water dropwort (*Oenanthe* sp.) roots to nematodes.

Davydkova et al. investigated the heavy metal content of polychaetes and free-living nematodes collected from benthos and bottom sediment in the Sea of Japan [135]. The number of investigated animals in case of different samples was not given exactly. It was only said that 2-100 animals were investigated depending on their size. In order to remove sediment particles from the outer part of the animals, they were washed with seawater. In order to purge the intestine, the washed worms were stored in a seawater-containing vessel for 1-4 hours. Acidic digestion was applied with concentrated HNO_3 for 24 hours at room temperature, then samples were heated to 120 °C. FAAS (acetylene-air flame) method was applied for elemental analysis. High concentration of iron, copper, zinc and cadmium was found in nematodes. Zeitoun-Ghandour et al. investigated the role of two *C. elegans* metallothioneins (CeMT-1 and CeMT-2) in terms of handling essential zinc and toxic cadmium [136]. Several analytical techniques were applied, among others ICP-AES. Wild-type and genetically modified strains were investigated as well. In the latter case, CeMT-1 and CeMT-2 and double knockout (KO) strains were obtained. Treatment with either $CdCl_2$ or $ZnCl_2$ solution was performed on L1 stage nematodes and lasting till the nematodes reached the pre-adult stage (L4). A high amount of nematodes was investigated with ICP-AES, after digestion in 1 N HNO_3 solution. For cadmium, equivalent concentration was found in the wild-type and the CeMT-1 KO strain, while the CeMT-2 KO strains had about twice higher concentration. Finally, for the double KO, a significant increase of the cadmium burden was found. Consequently, if CeMT-2 is expressed, CeMT-1 does not play an important role in the cadmium detoxification, however, when CeMT-2 is absent, its role is played by CeMT-1 but not superbly. For zinc, the role of the two isoforms was found to be different.

All KO strains accumulated significantly more zinc, than the wild-type, and for CeMT-1 a slightly higher zinc concentration was found, than for CeMT-2. Consequently, both isoforms are needed for maintaining zinc homeostasis, but CeMT-1 plays a more significant role. The zinc content was determined in case of the cadmium treated samples as well. For CeMT-1 KO a decrease of zinc content was found, consequently cadmium induced strongly CeMT-2 resulting in an enhanced excretion not only for cadmium but also for zinc. The CeMT-2 KO showed no change in zinc content, indicating that in this case the zinc homeostasis is unaffected, unlike for double KO strain, where a significant increase in zinc content was found.

Due to the relatively low mass of free-living nematodes, their individual analysis is a considerable challenge. Since in many cases only a low number of nematodes can be collected from contaminated areas, there is a need to analyze the individual worm. Savoly et al. developed a TXRF method for the elemental analysis of adult female individuals of the plant-feeding species *Xiphinema vuittenezi* [137]. Nematodes were starved in tap water for 5 days, in order to remove the gut content. The authors established that this time was not enough for the total removal, but longer starvation was impossible due to the relatively low lifespan of the investigated worms. After starvation or artificial treatments, the nematodes were rinsed in tap water for 1 minute. The fresh weight of the nematode individuals was estimated by measuring the length and greatest body diameter using the following formula:

$$W=(L*D^2)/(1.6*10^6) \qquad (1)$$

where W is the fresh weight in μg, L and D stand for the length and greatest body diameter of the worm in μm, respectively. Digestion of the worms was carried out directly on the quartz sample holder plate, adding 5 μl concentrated HNO_3 containing gallium as internal standard. After the drying step at 80 °C for 10 minutes, oxygen plasma ashing was applied in order to complete the mineralization of the organic matrix. Following this sample preparation, the concentration of 9 micro- and trace elements were determined in untreated worms. The same method was applied to study the copper uptake from copper sulfate solutions containing artificial soil solution in half of their volume. Beyond the dose-dependent copper uptake, zinc loss was found also in dose-dependent manner. Figure 11. shows the result of the measurements.

Howell measured the concentration of five heavy metals in two marine nematode species (*Enoplus brevis* and *Enoplus communis*) collected in three sampling sites with different levels of contamination at the north-east coast of

Britain [138]. Gut content of nematodes was removed by mild ultrasonication after narcotizing the animals. Individual animals were dried in vacuum desiccator at room temperature to constant weight, which was measured using an electronic balance. Digestion was carried out at 200 °C with nitric acid. Concentration of copper, zinc, lead and cadmium was measured by GF-AAS method, while mercury by CV-AAS technique. It was established that the concentration of the mentioned elements in the nematodes fall close to the values found for other marine organisms. The same researcher studied the uptake, tissue distribution and loss of copper and zinc in the same two species of nematodes originated from north-east coast of Britain [139]. *E. brevis* and *E. communis* were collected from unpolluted and *E. brevis* also from polluted sampling sites. Artificial solutions containing seawater and $ZnSO_4$ or $CuSO_4$ solutions were prepared as treating agents. After treatment, nematodes were rinsed several times in seawater and dissected into six regions as follows: pharynx, tail, gut, cuticle, remaining tissues (involving hypodermis, muscle layer, reproductive organs), eggs. Samples consisted of the above-mentioned parts of several nematodes. Digestion was carried out at 200 °C applying nitric acid.

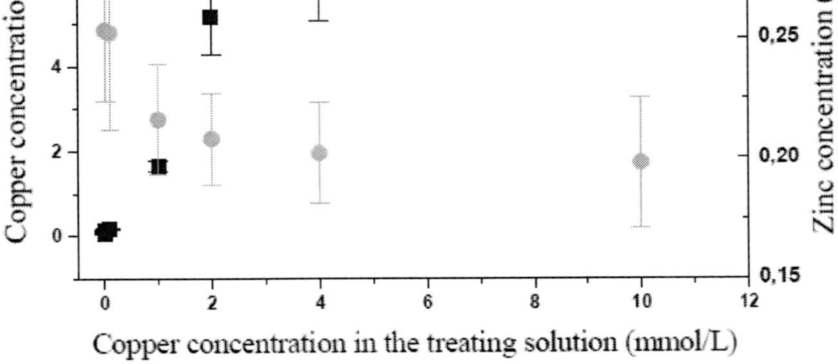

Figure 11. Effect of copper treatment on the copper and zinc concentration of the plant-feeding nematode Xiphinema vuittenezi.

Copper and zinc concentrations were determined by GF-AAS method. Results were expressed in ppm for time-course investigations, and accumulation factors (defined as tissue concentration after 12 days/tissue concentration after 0 days) were calculated as well. For all the three groups of nematodes a similar pattern was found concerning the copper uptake and loss of the cuticle: rapid uptake and immediate loss. Tail and pharynx showed similar characteristics probably due to the cuticular fraction of these tissues. Copper uptake of the gut was small for *E. brevis*, however, remarkable for *E. communis*. Other tissues showed similar differences: steady accumulation and no observable loss for the first and noticeable loss for the latter species. In case of zinc, less marked uptake was found for *E. brevis*, in contrast the highest accumulation factor was obtained for *E. communis* throughout the experiments. Significant gut uptake was found for the latter one and no uptake for the first one. It was established that the primary event is the surface absorption in all cases. In the same year, Howell also studied heavy metal uptake of mucus excreted by nematodes [140]. The same groups of nematodes were investigated as before [139]. A solution containing $CuSO_4$, $ZnSO_4$ and $Pb(NO_3)_2$, furthermore seawater was applied as a treating agent. Approximately 30 worms were placed into a vial containing the treating agent and acid-washed borosilicate powder for 30 minutes. This latter was added in order to collect the mucus on the particles. After removing the worms, the centrifugation was carried out in order to form a pellet containing the powder and the mucus. The pellet was washed in seawater, then the mucus was dissolved in 0.3 mL of acetic acid. This sample preparation step was repeated the same way without nematodes, in order to produce blank values. GF-AAS method was used for the measurement of copper, zinc and lead. It was impossible to determine the concentration values, because of the high water content of mucus. The measured intensities were compared to blank values and the significant differences proved the important role of mucus in the heavy metal binding.

Elemental Distribution Measurements

Jackson et al. investigated the distribution of copper and zinc in the body of *Caenorhabditis elegans* by means of SXRF method [141]. As treating agents, $CuCl_2*2H_2O$ or $Pb(NO_3)_2$ or both were dissolved in K-medium. After treatments, nematodes were preserved in formaldehyde and stored at 4 °C prior to the shipment.

The worms were mounted on Kapton tape in agar gel. The exciting X-ray radiation produced by the X-26A beamline of the National Synchrotron Light Source (Brookhaven, US) was tuned to 13 keV for Pb Lα and to 8.5 keV for Cu Kα excitation applying a Si(111) channel-cut monochromator. The beam was focused to an approximate size of 10 μm*10 μm using rhodium-coated Kirkpatrick-Baez focusing optics. Dwell time varied between 2 and 10 s. Copper distribution was quite homogeneous in the nematode, the lowest intensity was found at the external parts of the worm. In contrast, lead was found almost exclusively in the pharynx. This part of the nematode contains a high amount of neurons; thus, the findings are in agreement with the well-known neurotoxic effect caused by lead. At the applied beamline, X-ray diffraction measurements were also carried out. It was established that the crystal found in the pharynx is pyromorphite ($Pb_5(PO_4)_3Cl$). So the biomineralization of lead in the body of *C. elegans* has been proven. Gao et al. investigated copper distribution with the same technique on the same species [142]. Treatments were carried out on L1 stage worms for 36 hours using either $CuCl_2$ solutions or copper nanoparticles. For comparison, untreated nematodes were investigated as well. Nematodes were put onto a polycarbonate film fixed on plastic frame and quickly frozen in liquid nitrogen for a few seconds, and were kept at room temperature until the measurements. Elemental mapping was carried out at the BL-4A beamline of the Photon Factory (Japan). The energy of the incident beam was set to 13.5 keV using a synthetic multilayer monochromator (SMM). The beam was focused to an approximate size of 3 μm*5 μm using rhodium-coated Kirkpatrick-Baez focusing optics. Each point was measured for 10 s. For nanoparticles, just a slight uptake rate was observed. Increased copper levels were found in the head and at a location 1/3 of the way up the body from the tail. Much higher copper intensities were measured in case of treatment with $CuCl_2$ solution, especially in the excretory cells and intestine. It was found that due to copper uptake, the distribution of essential elements was changed. For example, the iron and zinc content of the proximal gonad increased for both treatments. Savoly et al. investigated the distribution of copper, iron and sulfur in the body of the plant-feeding nematode, *Xiphinema vuittenezi* treated with 1 mmol/L $CuSO_4$ solution for 24 hours [143]. The nematodes were put onto Kapton tape mounted on plastic holder. The samples were plunged into liquid nitrogen for 2 minutes in order to make quick-freezing, then lyophilized for 72 hours. The measurements were carried out at the BAMline beamline of the BESSY II synchrotron (Berlin, Germany). Elemental maps were collected using a 2 μm*2 μm spot size.

The beam energy was set to 9.7 keV using double multilayer monochromator. The measuring time was 1 s/point. The elemental maps are presented in Figure 12.

Elemental distribution was investigated in two areas. In case of iron, quite homogeneous distribution was found in both areas as well as for copper and sulfur in area 2. These two elements had similar pattern in the head region, increasing concentration toward the center of the investigated part. High correlation was found between the distribution of copper and sulfur in both areas.

Qu et al. investigated the fate of quantum dot nanoparticles taken up by *Caenorhabditis elegans* [144]. In case of CdSe@ZnS quantum dots, their distribution was investigated by the usual fluorescence imaging technique. Furthermore, elemental distribution measurements were also carried out using the SXRF technique at the BL-4A station of the Photon Factory. The energy of the incident beam was set to 13.5 keV by SMM, and it was focused to an approximate size of 5 μm*5 μm, using rhodium-coated Kirkpatrick-Baez optics. Fluorescent photons were collected for 1.5 s for each point.

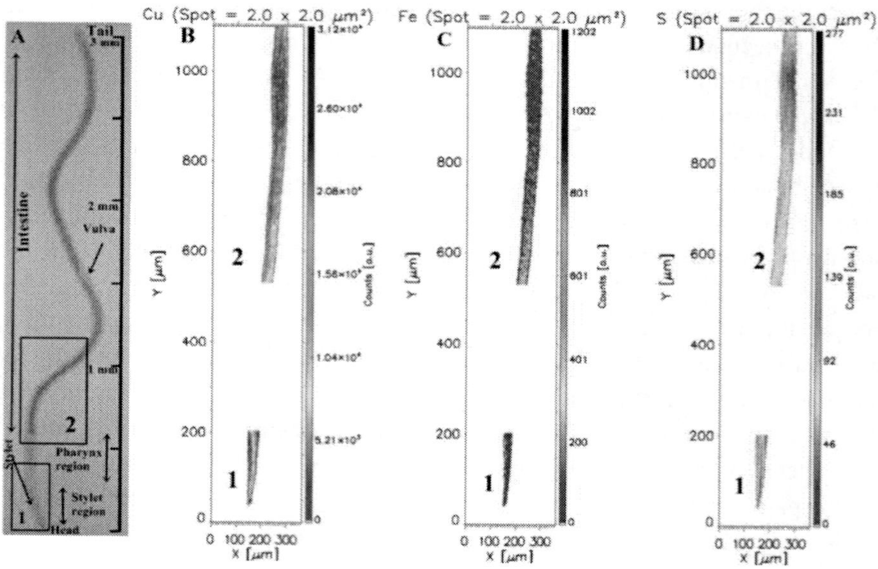

Figure 12. The body structure of a female nematode *Xiphinema vuittenezi* (A). Distribution of copper (B), iron (C) and sulfur (D) in the anterior part of another female nematode treated with 1 mmol/L $CuSO_4$ solution for 24 hours, 1: head, stylet and pharynx, 2: investigated part of the intestine.

In case of 24 h exposure time, no fluorescence signal was detected at the retral part of the intestine, however, with SXRF Se and Zn could be detected. It can be established that the SXRF method is more appropriate to investigate the nanoparticle distribution, than the fluorescence microscopy, because in case of chemical changes, fluorescence signal can be quenched, while the X-ray fluorescence signal is detectable. McColl et al. applied a tomographic method to investigate the three-dimensional distribution of essential metals in *Caenorhabditis elegans* [145]. After several washing steps, individuals were deposited onto a silicon nitride window and straightened by an eyelash. The window was frozen in liquid nitrogen chilled liquid propane and lyophilized overnight. Following the lyophilization, the straightest sample was removed and sandwiched by its tail between two pieces of adhesive tape. The energy of the applied beam was set to 10 keV. A spot with 1.5 μm diameter was applied prepared using a Kirkpatrick-Baez mirror pair. One hundred projections were obtained over 360°. Each projection was generated with an effective dwell time of 3 ms per pixel. One pixel had 1.6 μm^2 area. Measurements were carried out at the X-ray Fluorescence Microscopy beamline of the Australian Synchrotron. In the intestinal cells, an inhomogeneous distribution was found for both calcium and manganese. The intestinal cytoplasm was rich in iron, however, iron was excluded from the nuclei, unlike zinc, which concentrated there. The same group compared two kinds of samples: lyophilized (prepared similarly as in [145]) and hydrated (without water removal) [146]. The hydrated sample was covered with a 4 μm thick Ultralene film, to prevent evaporative loss. The measurements were carried out at the same beamline mentioned above. Incident X-ray energy was set to 12.73 keV and the spot size was 2 μm*2 μm (prepared by Kirkpatrick-Baez mirror pair). The dwell time was 2 ms per point, resulting in 18 minutes total acquisition time per specimen, which is a considerable short time (the above-mentioned measurements needed several hours or in certain cases several days). The stability of hydrated sample during the measurements was studied by investigating the re-distribution of elements. The authors establish that both sample preparation methods are suitable for the SXRF measurements.

Intracellular inclusions were studied in nematodes inhabiting thiobiotic mangrove mud-flats with SEM, TEM and EPMA methods by Nicholas et al. [147]. The individuals of three nematode species (*Sabatieria wieseri*, *Terschellingia longicaudata*, *Sphaerolaimus papillatus*) were investigated. SEM-EPMA system was applied for semiquantitative analysis of thick sections. TEM combined with EPMA was used to localize elements within the inclusions.

For TEM study, the following sample preparation steps were used: fixation with phosphate buffered 3% glutaraldehyde; washing in phosphate buffered 2% sucrose; post-fixation in 1 % osmium tetroxide in phosphate buffer; repeated washing with water, transfer in steps to ethanol, propylene oxide and Spurr resin. After hardening in the epoxy resin, silver sections were cut with glass knife and stained on copper grids with Reynold's lead citrate and uranyl acetate. In case of SEM and TEM-EPMA investigations, similar sample preparation method was applied, however, in these cases treatment with osmium tetroxide, lead citrate, uranyl acetate and buffers were omitted in order to avoid any contamination. For SEM, 5 μm sections were cut and mounted onto carbon stubs. In case of TEM-EPMA analysis, the copper grids were coated with carbon to provide stability. According to the SEM-EPMA measurements on thick epoxy sections, in tissues silicon, phosphorus, sulfur, potassium, calcium and iron were detected. The intracellular inclusions were investigated by TEM-EPMA, beyond the above-mentioned elements, they contain sodium, zinc and aluminum. As the authors observed, it is possible that metal sulfides are deposited on a calcium pyrophosphate core in order to prevent harmful accumulation.

Maybe metallothioneins are also involved. No proof was found pro or contra. Williams and Seraphin investigated lead biomineralization in the bacterial feeder rhabditid *Panagrolaimus superbus* [148]. The worms were treated with lead acetate solution containing lead in concentration of 500 ppm for 20 minutes. After washing with deionized water, the nematodes were immersed into 2% glutaraldehyde buffered with 0.1 M phosphate for 24 h. Dehydration was carried out through an alcohol series and the animals were embedded in Spurr resin. Sections cut by microtome were stained in order to obtain TEM micrographs, while in case of EPMA measurements no staining was applied. Particulates were found in the semicular canal lumen and the oesophagal gland tissue. The granules contained lead, phosphorus, sulfur and calcium in both cases. In contrast, in case of the vicinity of the granules, these elements were not detected. It is likely that similar processes took place also as in case of *C. elegans* investigated by SXRF [141]. Savoly et al. studied the elemental distributions in *Xiphinema vuittenezi* individuals using SEM, FIB and EPMA methods [149]. The sample preparation was similar as discussed in [143]. Quick- freezing and lyophilization were applied, in this case the nematodes were mounted onto Carbon Adhesive Tape developed for electron microscopic applications. Lyophilized and non-lyophilized nematodes were bombarded with Ga^+ ions using the FIB unit. Figure 13. compares the effect of ion bombardment on non-lyophilized and lyophilized samples.

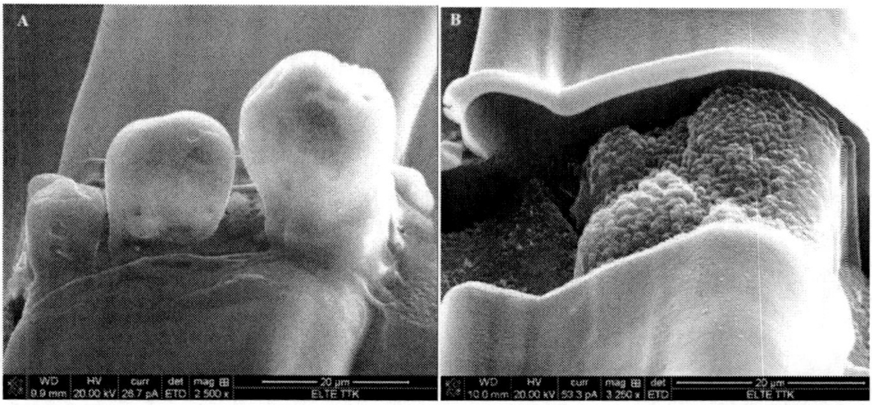

Figure 13. The effect of Ga^+ beam on non-lyophilized (A) and lyophilized (B) nematode samples.

It was found that from lyophilized samples appropriate cross-sections can be prepared by FIB, while the non-lyophilized worm was destroyed. Copper content of the head region was investigated by EPMA technique (Figure 14).

It should be mentioned that due to the low analytical depth of the EPMA method, only the external regions (the cuticle) and the stylet (if protruded) could be investigated.

The sulfur content of the lip region and the stylet was significantly higher than in the cuticle of the head. Elevated copper concentration was found in the lips, while the copper content of the stylet was quite low.

In order to study the elemental distribution within the nematodes, FIB technique was used. Artificial cross-sections were prepared in case of three nematodes treated with 1 mmol/L $Cu(NO_3)_2$, 50 mmol/L K_2CrO_4 solution or 50 mg/L nano ZnO suspension for 24 h. In the cross-sections, one-dimensional elemental distribution measurements were carried out. Figure 15. demonstrates the distribution of copper, phosphorus and sulfur in the cross-section. Correlation between the distribution of copper, chromium, zinc and sulfur and phosphorus was investigated. High correlation was found for Cr-P, Cr-S and Cu-S and moderate correlation for all other combinations.

In all cases, similar pattern was found: decreasing intensity toward the center of the cross-section.

Consequently, it is likely for all the three metal contaminants that the cuticular uptake plays a prominent role. The contradiction between these findings and [141] may be due to either the difference in the applied techniques, or the considerably different taxon applied.

The SXRF method measures the whole volume in a certain irradiated point, and due to the fact that the shape of the nematode is not perfectly cylindrical, the element content of the external areas is underestimated, since the investigated volume is smaller here.

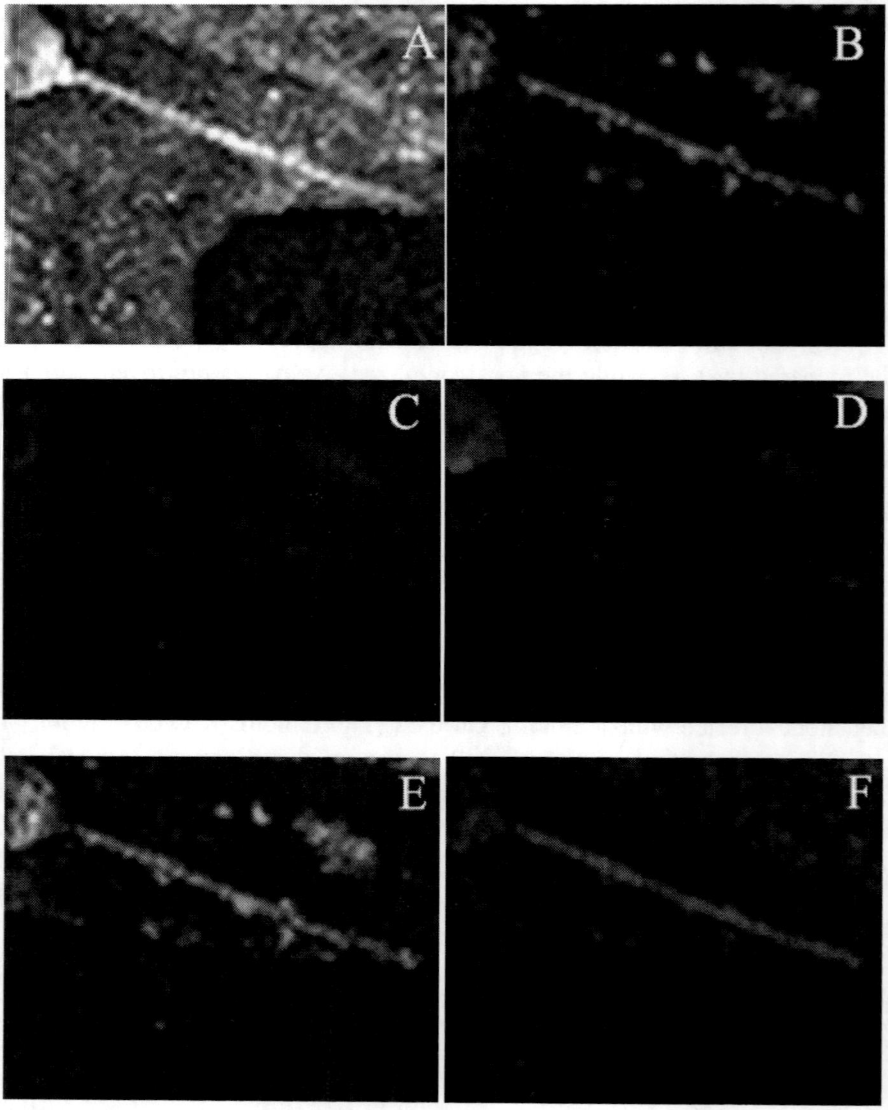

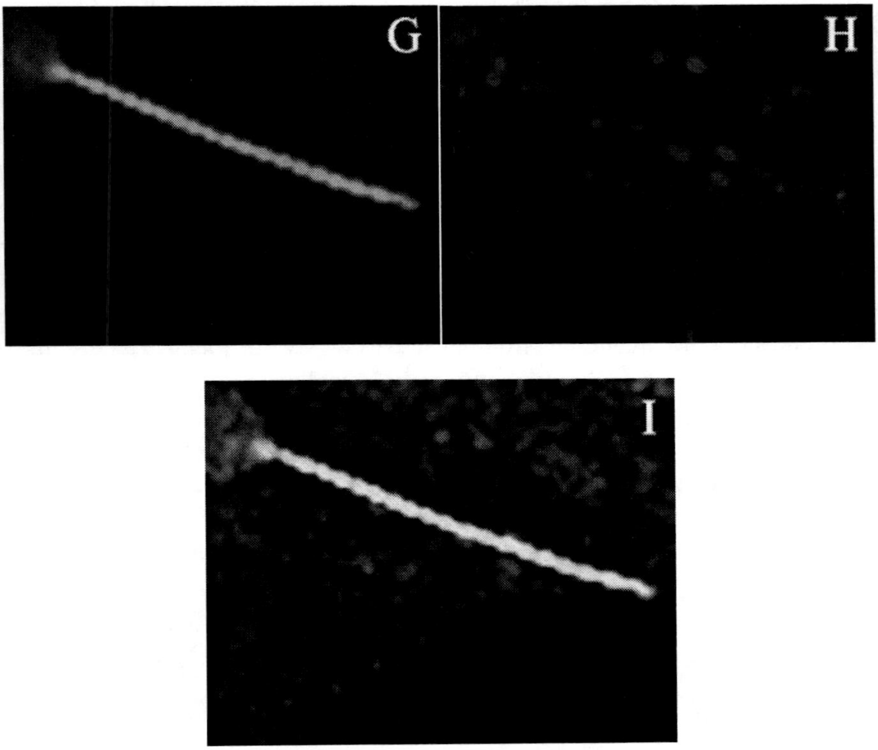

Figure 14. Elemental distribution maps measured in the stylet and head region of the nematode treated with 1 mmol/L $CuSO_4$ solution. Secondary electron picture (A), distribution of nitrogen (B), oxygen (C), phosphorus (D), sum of pictures B, C and D (E), sodium (F), sulfur (G), copper (H), sum of pictures F, G and H (I).

Another possible source for the observed differences can be the taxonomic discrepancy between the studied species. While *Caenorhabditis elegans* is a bacterial feeding nematode belonging to order Rhabditida of the class Secernentia, known to be relatively stress-tolerant [150], *Xiphinema vuittenezi* is a member of order Dorylaimida of the class Penetrantia. Dorylaimid nematodes are known to be rather sensitive to chemical pollution effects, due to the permeability of their cuticle [150].

These differences underline the importance of involving species of as many taxonomic groups and life strategies as possible into these types of studies.

Leapman et al. applied electron tomography with EELS to investigate three-dimensional phosphorus distribution in *C. elegans* [151]. For sample preparation, the freeze-substitution technique was applied.

At first, high pressure freezing was carried out and then freeze-substitution using acetone containing 0.2 % ethylene glycol-bis-succinimidil succinate.

The latter process has taken 3 days, then specimens were gradually warmed up to room temperature and embedded into Epon-Araldite by graded exchange of the acetone. Specimens were polymerized and sectioned by an ultramicrotome to a thickness of about 70 nm. Sections were mounted onto copper grids. On the 3D distribution maps, small objects with diameter of 15-20 nm were observed and identified as ribosomes. Cranfield et al. detected biogenic magnetite in *Caenorhabditis elegans* [152]. Nematodes were freeze-dried. On the TEM image electron-dense particles were found. According to the EPMA and EELS measurements, it was established that particles contain only iron and oxygen and are surrounded by calcium-rich crystalline matrix. The presence of magnetite may explain the production of heat-shock proteins in response to mobile phone-type electromagnetic field exposure.

Investigation of Elemental Speciation

Zeitoun-Ghandour et al. conducted XANES and EXAFS measurements as well [136]. Samples for XAS were prepared from high amount of nematodes (300,000-500,000).

A

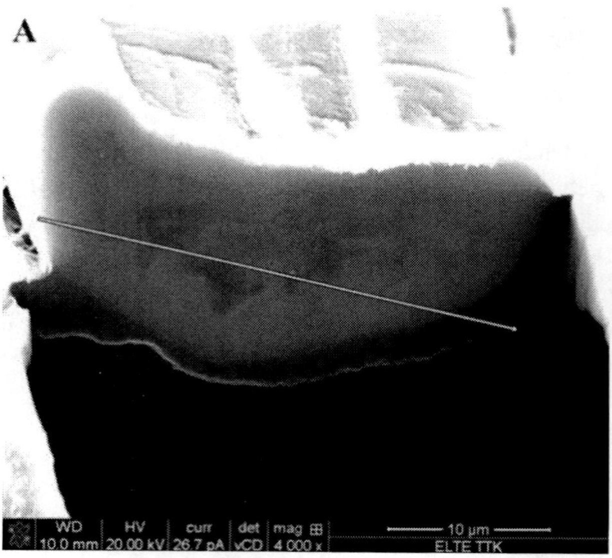

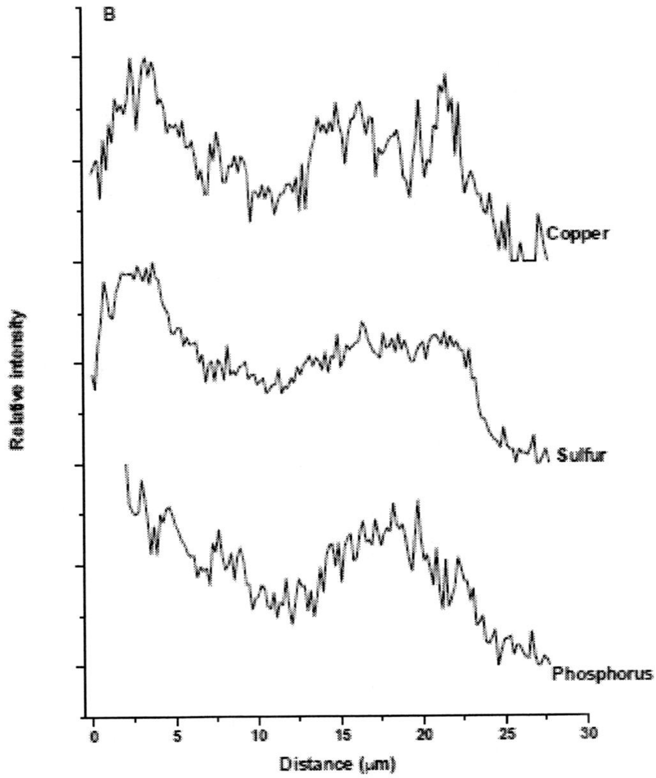

Figure 15. A: Cross-section of the nematode treated with 1 mmol/L $Cu(NO_3)_2$ solution prepared near to the tail. B: Distribution of phosphorus, sulfur and copper along the marked line.

Fine powder was generated from nematodes under liquid nitrogen and stored at -80 °C. XANES and EXAFS spectra were obtained at the K-edge of cadmium or zinc. Measurements were carried out on the station 16.5 of the Synchrotron Radiation Source (Warrington, UK). Samples were mounted onto aluminum sample holders and the XAS measurements were carried out at cryogenic temperature (approximately 20 K). Standard samples were grinded in agate mortar and diluted with boron nitride. Transmission mode was applied. The cadmium XANES and EXAFS measurements show that this element is sulfur-coordinated. On the one hand, XANES spectrum of the wild-type *C. elegans* is similar to the spectra of the sulfur-coordinated model compounds. On the other hand, the best EXAFS model was found in case of four sulfur scatterers at a distance of 2,94 Å from the cadmium absorber.

In contrast, zinc was found to be bound by phosphorus-containing ligands. Studying the XANES spectrum of the zinc-treated wild-type nematode, the highest similarity was found for $Zn_3(PO_4)_2$ spectrum.

Furthermore, the best fit for EXAFS spectrum was found for 4 oxygen scatterers at a distance of 1.97 Å. The EXAFS and XANES spectra were similar for the double KO. Cadmium is not bound to metallothionein, since in case of the double KO, cadmium was bound mainly by sulfur as well. Consequently, another sulfur-containing ligand has to take part in cadmium binding. It is likely that phytochelatins play a dominant role in cadmium detoxification.

As mentioned above, Qu et al. applied XANES spectrometry to study the fate of quantum dots in the body of *C. elegans* [144]. The spot size was the same as in case of SXRF measurements, XANES spectra were obtained at Se K-edge in fluorescence mode. The XANES spectrum in the intestine was similar to the spectrum of the quantum dots, however, in its retral part and in hindgut was found to take place as SeO_3^{2-}.

Consequently, due to the digestion, the core is collapsed and Se^{2-} is oxidized to SeO_3^{2-} in the body of the nematode. This process caused the quench of fluorescent signal as mentioned earlier.

Savoly et al. investigated the chemical environment of copper in the body of *Xiphinema vuittenezi*. The midbody of the nematode was investigated with a spot size of about 300 μm*1350 μm. Solid reference compounds were also studied. Several treatments were applied and an untreated nematode was investigated as well. It was established that copper occurs in the body of the nematodes in all cases mainly as Cu^{2+}.

According to the XANES spectra (Figure 16.), it can be stated that the different parameters of the treatments (concentration and anion of the treating agent, time of treatment) had not significant effect on the chemical environment of the copper taken up from the solutions. However, the spectrum was quite different for the untreated worm. Thus, the essential copper and the copper taken up by artificial treatments occur in different forms.

CONCLUSION

Nematodes are often studied in different scientific areas. They are applied in environmental science as bioindicators, and certain nematode species, in particular the bacterial feeder *Caenorhabditis elegans*, are often used as model organisms in biological and biochemical investigations.

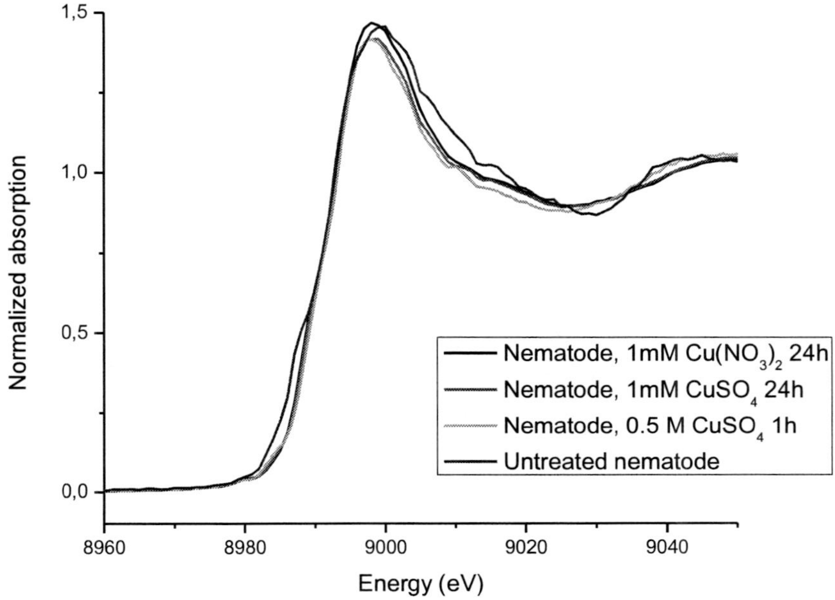

Figure 16. XANES spectrum of copper-treated and untreated nematode samples.

However, most of the other nematode feeding types are hardly or not at all represented among the studied species. Furthermore, several species deserve considerable attention as important crop pests. Therefore the study of metal uptake and release characteristics of nematodes has fundamental importance to understand the biochemical processes generating stress responses. Using modern instrumental microanalytical techniques, the application range of nematodes can be significantly broadened. Through a few samples this review demonstrates that more exact chemical information can be provided on microscopic living systems helping us to understand different biochemical processes and to explain their stress responses on different contaminants. Unfortunately, most of the analytical systems are expensive; however, their use is inevitable. It is enough to think of the difference between the traditional fluorescence microscopy and X-ray spectroscopic techniques. Due to the technological developments, the detection limits are becoming lower and the resolution in 3D higher and higher. The resolutions are getting better for both elemental distribution – studies on the cell level are allowed recently –, and XAS measurements. This latter will result in more exact characterization of the chemical environment of toxic micro- or trace elements.

Consequently, our nematological knowledge will increase from several important aspects. For example, in crop protection, antagonistic effects of some metals, for instance copper used as a pesticide can be studied on economically important phytonematodes, e.g. the *Xiphinema* species referred often in this work. Alternatively, exploring the contaminant uptake and release patterns will become possible even for those, ecologically important taxa (e.g. predatory nematodes belonging to the order Dorylaimida), whose lower density values and difficult or impossible mass culturing have made them unavailable for this type of studies so far. This way, our knowledge will become more complete on this field. Thanks to this, more convincing explanations can be expected for the mechanisms of toxicity in this important regulator and bioindicator group in terrestrial systems.

References

[1] Azmat, R.; Fayyaz, S.; Kazi, N.; Mahmood, S. J.; Uddin, F. (2008). Natural bioremediation of heavy metals through nematode parasite of fish. *Biotechnology*, 7, 139-143.

[2] Nagy, P.; Bakonyi, G.; Bongers, T.; Kadar, I.; Fabian, M.; Kiss, I. (2004). Effects of microelements on soil nematode assemblages seven years after contaminating an agricultural field. *Science of the Total Environment,* 320, 131-143.

[3] Palikova, M.; Barus, V. (2003). Mercury content in *Anguillicola crassus* (Nematoda) and its host *Anguilla anguilla*. *Acta Veterinaria Brno*, 72, 289-294.

[4] Bongers, T.; Ferris, H. (1999). Nematode community structure as a bioindicator in environmental monitoring. *Trends in Ecology and Evolution,* 14, 224-228.

[5] Horvitz, H. (2003). Worms, life, and the death (Nobel lecture). *ChemBioChem*, 4, 697-711.

[6] Tatar, M.; Bartke, A.; Antebi, A. (2003). The endocrine regulation of aging by insulin-like signals. *Science*, 299, 1346-1351.

[7] Kurfurst, U. *Solid Sample Analysis*. Berlin: Springer; 1998.

[8] Ertas, O. S.; Tezel, H. (2004). A validated cold vapour-AAS method for determining mercury in human red blood cells. *Journal of Pharmaceutical and Biomedical Analysis*, 36, 893-897.

[9] De Loos-Vollebregt, M. T. C. (2013). Background correction methods in atomic absorption spectrometry. In: Meyers, R. A. (Ed.), *Encyclopedia*

of Analytical Chemistry (pp. 1-17.) New York: John Wiley and Sons Ltd.

[10] Welz, B.; Becker-Ross, H.; Florek, S.; Heitmann, U. *High-resolution continuum source AAS – the better way to do atomic absorption spectrometry*. Weinheim: Wiley-VCH; 2005.

[11] Locke, J. (1979). The determination of eight elements in human liver tissue by flame atomic absorption spectrometry in sulphuric acid solution. *Analytica Chimica Acta*, 104, 225-231.

[12] Tinggi, U.; Maher, W. (1986). Determination of trace elements in biological tissues by aluminum block digestion and spike-height flame atomic absorption spectrometry. *Microchemical Journal*, 33, 304-308.

[13] Murphy, V. A. (1987). Method for determination of sodium, potassium, calcium, magnesium, chloride, and phosphate in the rat choroid plexus by flame atomic absorption and visible spectroscopy. *Analytical Biochemistry,* 161, 144-151.

[14] Nunes, L. S.; Barbosa, J. T. P.; Fernandes, A. P.; Lemos, V. A.; dos Santos, W. N. L.; Korn, M. G. A.; Teixeira, L. S. G. (2011). Multi-element determination of Cu, Fe, Ni and Zn content in vegetable oils samples by high-resolution continuum source atomic absorption spectrometry and microemulsion sample preparation. *Food Chemistry*, 127, 780-783.

[15] Bechlin, M. A.; Neto, J. A. G.; Nobrega, J. A. (2013). Evaluation of lines of boron, phosphorus and sulfur by high-resolution continuum source flame atomic absorption spectrometry for plant analysis. *Microchemical Journal*, 109, 134-138.

[16] Ati-Hellal, M. E.; Hellal, F.; Dachraoui, M.; Hedhili, A. (2007). Plackett-Burman designs in the pretreatment of macroalgae for Pb, Cr and Al determination by GF-AAS. *Comptes Rendus Chimie*, 10, 839-849.

[17] De Sousa, R. A.; Sabarense, C. M.; Prado, G. L. P.; Metze, K.; Cadore, S. (2013). Lead biomonitoring in different organs of lead intoxicated rats employing GF AAS and different sample preparations. *Talanta*, 104, 90-96.

[18] Ribeiro, A. S.; Vieira, M. A.; da Silva, A. F.; Borges, D. L. G.; Welz, B.; Heitmann, U.; Curtius, A. J. (2005). Determination of cobalt in biological samples by line-source and high-resolution continuum source graphite furnace atomic absorption spectrometry using solid sampling or alkaline treatment. *Spectrochimica Acta B*, 60, 693-698.

[19] Borges, D. L. G.; da Silva, A. F.; Welz, B.; Curtius, A. J.; Heitmann, U. (2006). Determination of lead in biological samples by high resolution continuum source graphite furnace atomic absorption spectrometry with direct solid sampling. *Journal of Analytical Atomic Spectrometry*, 21, 763-769.

[20] Wendt, R. H.; Fassel, V. A. (1965). Induction-coupled plasma spectrometric excitation source. *Analytical Chemistry*, 37, 920-922.

[21] Tormen, L.; Gil, R. A.; Frescura, V. L. A.; Martinez, L. D.; Curtius, A. J. (2010). Determination of trace elements in biological samples treated with formic acid by inductively coupled plasma mass spectrometry using a microconcentric nebulizer. *Spectrochimica Acta* B, 65, 959-966.

[22] Kretschy, D.; Groger, M.; Zinkl, D.; Petzelbauer, P.; Koellensperger, G.; Hann, S. (2011). High-throughput flow injection analysis of labeled peptides in cellular samples – ICP-MS analysis versus fluorescence based detection, *International Journal of Mass Spectrometry*, 307, 105-111.

[23] Kantor, T.; Zaray, G. (1995). Improved design and optimization of an electrothermal vaporization system for inductively coupled plasma atomic emission spectrometry, *Microchemical Journal,* 51, 266-277.

[24] Tormen, L.; Gil, R. A.; Frescura, V. L. A.; Martinez, L. D.; Curtius, A. J. (2012). The use of electrothermal vaporizer coupled to the inductively coupled plasma mass spectrometry for the determination of arsenic, selenium and transition metals in biological samples treated with formic acid, *Analytica Chimica Acta*, 717, 21-27.

[25] Zaray, G.; Kantor, T. (1995). Direct determination of arsenic, cadmium, lead and zinc in soils and sediments by electrothermal vaporization and inductively coupled plasma excitation spectrometry. *Spectrochimica Acta B*, 50, 489-500.

[26] Vanhoe, H.; Goossens, J.; Moens, L.; Dams, R. (1994). Spectral interferences encountered in the analysis of biological materials by inductively coupled plasma mass spectrometry. *Journal of Analytical Atomic Spectrometry*, 9, 177-185.

[27] Cindric, I. J.; Krizman, I.; Zeiner, M.; Kampic, S.; Medunic, G. (2012). ICP-AES determination of minor- and major elements in apples after microwave assisted digestion. *Food Chemistry*, 135, 2675-2680.

[28] Cindric, I. J.; Zeiner, M.; Krpetic, M.; Stingeder, G. (2012). ICP-AES determination of minor and major elements in Cornelian cherry (*Cornus mas* L.) after microwave assisted digestion. *Microchemical Journal*, 105, 72-76.

[29] Sivakumar, S.; Khatiwada, C. P.; Sivasubramanian, J. (2012). Bioaccumulations of aluminum and the effects of chelating agents on different organs of *Cirrhinus mrigala*. *Environmental Toxicology and Pharmacology,* 34, 791-800.

[30] Tokalioglu, S. (2012). Determination of trace elements in commonly consumed medicinal herbs by ICP-MS and multivariate analysis. *Food Chemistry,* 134, 2504-2508.

[31] Dressler, V. L.; Pozebon, D.; Matusch, A.; Becker, J. S. (2007). Micronebulization for trace analysis of lanthanides in small biological specimens by ICP-MS. *International Journal of Mass Spectrometry*, 266, 25-33.

[32] Llorent-Martinez, E. J.; de Cordova, M. L. F.; Ruiz-Medina, A.; Ortega-Barrales, P. (2012). Analysis of 20 trace and minor elements in soy and dairy yogurts by ICP-MS. *Microchemical Journal*, 102, 23-27.

[33] Dressler, V. L.; Antes, F. G.; Moreira, C. M.; Pozebon, D.; Duarte, F. A. (2011). As, Hg, I, Sb, Se and Sn speciation in body fluids and biological tissues using hyphenated-ICP-MS techniques: A review. *International Journal of Mass Spectrometry*, 307, 149-162.

[34] Klockenkamper, R. *Total Reflection X-Ray Fluorescence Analysis*. New York: John Wiley and Sons Ltd.; 1997.

[35] Moseley, H. G. J. (1913, 1914). The high-frequency spectra of the elements. *Philosophical Magazine*, 26, 1024-1034., 27, 703-713.

[36] Yoneda, Y.; Horiuchi, T.; Optical flats for use in X-ray spectrochemical microanalysis. *Review of Scientific Instruments*, 42, 169-170.

[37] Compton, A. H. (1923). The total reflection of X-rays. *Philosophical Magazine,* 45, 1121-1131.

[38] Injuk, J.; van Grieken, R.; Klockenkamper, R.; von Bohlen, A.; Kump, P. (1997). Performance and characteristics of two total-reflection X-ray fluorescence and a particle induced X-ray emission setup for aerosol analysis. *Spectrochimica Acta B,* 52, 977-984.

[39] Huhnerfuss, K.; von Bohlen, A.; Kurth, D. (2006). Characterization of pigments and colors used in ancient Egyptian boat models. *Spectrochimica Acta B*, 61, 1224-1228.

[40] Von Bohlen, A.; Klockenkamper, R.; Tolg, G.; Wiecken, B. (1988). Microtome sections of biomaterials for trace analyses by TXRF. *Fresenius Journal of Analytical Chemistry*, 331, 454-458.

[41] Mages, M.; Woelfl, S.; Ovari, M.; von Tumpling, W.; Encina, F. (2004). The use of a portable total reflection X-ray fluorescence spectrometer for

trace element determination in freshwater microcrustaceans (*Daphnia*). *Spectrochimica Acta B*, 59, 1265-1272.

[42] Hoefler, H.; Streli, C.; Wobrauschek, P.; Ovari, M.; Zaray, G. (2006). Analysis of low Z elements in various environmental samples with total reflection X-ray fluorescence (TXRF) spectrometry. *Spectrochimica Acta B*, 61, 1135-1140.

[43] Szoboszlai, N.; Polgari, Z.; Mihucz, V. G.; Zaray, G. (2009). Recent trends in total reflection X-ray fluorescence spectrometry for biological applications. *Analytica Chimica Acta*, 633, 1-18.

[44] Barthelme, D.; Scheele, U.; Dinkelaker, S.; Janoschka, A.; MacMillan, F.; Albers, S.-V., Driessen, A. J. M.; Stagni, M. S.; Bill, E.; Meyer-Klaucke, W.; Schünemann, V.; Tampe, R. (2007). Structural organization of essential iron-sulfur clusters in the evolutionarily highly conserved ATP-binding cassette protein ABCE1; *Journal of Biological Chemistry,* 282, 14598-14607.

[45] Schilling, S.; Lindner, C.; Koch, B.; Wermann, M.; Rahfeld, J.-U.; von Bohlen, A.; Rudolph, T.; Reuter, G.; Demuth, H.-U. (2007). Isolation and characterization of glutaminyl cyclases from *Drosophila*: evidence for enzyme forms with different subcellular localization. *Biochemistry*, 46, 10921-10930.

[46] Ruiz, R. F.; Tornero, J. D.; Gonzalez, V. M.; Alonso, C. (1999). Quantification of Pt bound to DNA using total-reflection X-ray fluorescence (TXRF). *Analyst*, 124, 583-585.

[47] Messerschmidt, J.; von Bohlen, A.; Alt, F.; Klockenkamper, R. (2000). Separation and enrichment of palladium and gold in biological and environmental samples, adapted to the determination by total reflection X-ray fluorescence. *Analyst*, 125, 397-399.

[48] Greaves, E. D.; Parra, L. M. M.; Rojas, A.; Sajo-Bohus, L. (2000). Determination of platinum levels in serum and urine samples from pediatric cancer patients by TXRF. *X-ray Spectormetry*, 29, 349-353.

[49] Prange, A.; Bodekker, H.; Michaelis, W. (1989). Multi-element determination of trace elements in whole blood and blood serum by TXRF. *Fresenius Journal of Analytical Chemistry*, 335, 914-918.

[50] Yap, C. T. (1988). X-ray total reflection fluorescence analysis of iron, copper, zinc, and bromine in human serum. *Applied Spectroscopy*, 42, 1250-1253.

[51] Zarkadas, C.; Karydas, A. G.; Paradellis, T. (2001). Applicability of direct total reflection X-ray fluorescence analysis in the case of human blood serum samples. *Spectrochimica Acta B*, 56, 2219-2228.

[52] Carvalho, M. L., Custodio, P. J.; Reus, U. (2001). Elemental analysis of human amniotic fluid and placenta by total-reflection X-ray fluorescence and energy-dispersive X-ray fluorescence: child weight and maternal age dependence. *Spectrochimica Acta B*, 56, 2175-2180.

[53] Parra, L. M. M.; Hernandez-Caraballo, E. A. (2004). Direct analysis of biological samples by total reflection X-ray fluorescence. *Spectrochimica Acta B*, 59, 1077-1090.

[54] Ostachowicz, B.; Lankosz, M.; Tomik, B.; Adamek, D. Wobrauschek, P.; Streli, C.; Kregsamer, P. (2006). Analysis of some chosen elements of cerebrospinal fluid and serum in amyotrophic lateral sclerosis patients by total reflection X-ray fluorescence. *Spectrochimica Acta B*, 61, 1210-1213.

[55] Xie, M.; von Bohlen, A.; Klockenkamper, R.; Jian, X.; Günther, K. (1998). Multielement analysis of Chinese tea (*Camellia sinensis*) by total-reflection X-ray fluorescence. *Zetschrift für Lebensmittel-Untersuchung und Forschung*, 207, 31-38.

[56] Zaray, G.; Varga, A.; Fodor, F.; Cseh, E. (1997). Microanalytical investigation of xylem sap of cucumber by total reflection X-ray fluorescence spectrometry. *Microchemical Journal*, 55, 64-71.

[57] Arredondo, M.; Cambiazo, V.; Tapia, L.; Gonzalez-Aguero, M.; Nunez, M. T.; Uauy, R.; Gonzalez, M. (2004). Copper overload affects copper and iron metabolism in Hep-G2 cells. *American Journal of Physiology – Gastrointestinal and Liver Physiology*, 287, G27-G32.

[58] Szoboszlai, N.; Reti, A.; Budai, B.; Szabo, Z.; Kralovanszky, J.; Zaray, G. (2008). Direct elemental analysis of cancer cell lines by total reflection X-ray fluorescence. *Spectrochimica Acta B*, 63, 1480-1484.

[59] Espinoza-Quinones, F. R.; Modenes, A. N.; Palacio, S. M.; Szymanski, N.; Welter, R. A.; Rizzutto, M. A., Borba, C. E.; Kroumov, A. D. (2010). Evaluation of trace element levels in muscles, liver and gonad of fish species from Sao Francisco River of the Parana Brazilian state by using SR-TXRF technique. *Applied Radiation and Isotopes*, 68, 2202-2207.

[60] De la Calle, I.; Costas, M.; Cabaleiro, N.; Lavilla, I.; Bendicho, C. (2013). Fast method for multielemental analysis of plants and discrimination according to the anatomical part by total reflection X-ray fluorescence spectrometry. *Food Chemistry,* 138, 234-241.

[61] Magalhaes, T.; Carvalho, M. L.; von Bohlen, A.; Becker, M. (2010). Study on trace elements behaviour in cancerous and healthy tissues of

colon, breast and stomach: Total reflection X-ray fluorescence applications. *Spectrochimica Acta B*, 65, 493-498.

[62] Pathan, A. K.; Bond, J.; Gaskin, R. E. (2008). Sample preparation for scanning electron microscopy of plant surfaces – Horses for courses. *Micron*. 39, 1049-1061.

[63] Valdecasas, A. G.; Camacho, A. I. (2005). On the environmental scanning electron microscope for taxonomic purposes. *Invertebrate Biology,* 124, 66-73.

[64] Read, N. D.; Jeffree, C. E. (1991). Low-temperature scanning electron microscopy in biology. *Journal of Microscopy*, 161, 59-72.

[65] Anderson, T. F. (1951). Techniques for the preservation of three dimensional structure in preparing specimens for the electron microscope. *Transactions of the New York Academy of Sciences*, 13, 130-134.

[66] Boyde, A.; Barber, V. C. (1969). Freeze-drying methods for the scanning electron-microscopical study of the protozoon *Spirostomum ambiguum* and the statocyst of the cephalopod mollusc *Loligo vulgaris*. *Journal of Cell Science*, 4, 223-239.

[67] Hess, W. M. (1966). Fixation and staining of fungus hyphae and host plant root tissues for electron microscopy. *Stain Technology*, 41, 27-35.

[68] Don Ardenne, M.; Beischer, D. (1940). Untersuchung von Metalloxyd-Rauchen mit dem Universal-Elektronenmikroskop. *Zeitschrift für Elektrochemie und Angewandte Physikalische Chemie,* 46, 270-277.

[69] Knoll, M.; Ruska, E. (1932). Das Elektronenmikroskop. *Zeitschrift für Physik,* 78, 318-339.

[70] Chourasia, A. R.; Chopra, D. R. (1997). Auger Electron Spectroscopy. In: Settle, F. A. (Ed.), *Handbook of Instrumental Techniques for Analytical Chemistry* (pp. 791-808.) Upper Saddle River: Prentice Hall PTR.

[71] Egerton, R. F. (2009). Electron energy-loss spectroscopy in the TEM. *Reports on Progress in Physics*, 72, 016502 (25 pp).

[72] Castaing, R. (1951). Application of electron probes to local chemical and crystallographic analysis. *PhD Thesis*. Paris: University of Paris.

[73] Ortega, R. (2005). Chemical elements distribution in cells. *Nuclear Instruments and Methods in Physics Research B*, 231, 218-223.

[74] Swanson, L. W. (1983). Liquid metal sources: mechanism and applications. *Nuclear Instruments and Methods in Physics Research*, 218, 347-353.

[75] Volkert, C. A.; Minor, A. M. (2007). Focused ion beam microscopy and micromachining. *MRS Bulletin*, 32, 389-399.

[76] Langford, R. M., Nellen, P. M., Gierak, J., Fu, Y. (2007). Focused ion beam micro- and nano engineering. *MRS Bulletin*, 32, 417-423.

[77] Mayer, J.; Gianuzzi, L. A.; Kamino, T.; Michael, J. (2007). TEM sample preparation and FIB-induced damage. *MRS Bulletin*, 32, 400-407.

[78] Fujita, J.; Ishida, M.; Ichihashi, T.; Ochiai, Y.; Kaito, T.; Matsui, S. (2003). Growth of three-dimensional nano-structures using FIB-CVD and its mechanical properties. *Nuclear Instruments and Methods in Physics Research B*, 206, 472-477.

[79] Tanaka, K. (1989) High resolution electron microscopy of the cell. *Biology of the Cell*, 65, 89-98.

[80] Friedmann, A.; Hoess, A.; Cismak, A.; Heilmann, A. (2011) Investigation of cell-substrate interactions by focused ion beam preparation and scanning electron microscopy. *Acta Biomaterialia*, 7, 2499-2507.

[81] Scott, K. (2011). 3D elemental and structural analysis of biological specimens using electrons and ions. *Journal of Microscopy*, 7, 2499-2507.

[82] Smart, K. E.; Kilburn, M. R.; Salter, C. J.; Smith, J. A. C., Grovenor, C. R. M., NanoSIMS amd EPMA analysis of nickel localization in leaves of the hyperaccumulator plant *Alyssum lesbiacum*. *International Journal of Mass Spectrometry,* 260, 107-114.

[83] Franco, A.; Rufo, L.; de la Fuente, V. (2012). Metal concentration and distribution in plant tissues of *Nerium oleander* (Apocynaceae, Plantae) from extremely acidic and less extremely acidic water courses in the Rio Tinto area (Huelva, Spain). *Ecological Engineering*, 47, 87-91.

[84] Arru. L.; Rognoni, S.; Baroncini, M.; Bonatti, P. M.; Perata, P. (2004). Copper localization in Cannabis sativa L. grown in a copper-rich solution. *Euphytica*, 140, 33-38.

[85] Bassi, D.; Cappa, F.; Cocconcelli, P. S. (2009). A combination of a SEM technique and X-ray microanalysis for studying the spore germination process of *Clostridium tyrobutyricum*. *Research in Microbiology*, 160, 322-329.

[86] Al-Bahry, S. N.; Mahmoud, I. Y.; Al-Amari, I. S.; Ba-Omar, T. A.; Melgheit, K. O.; Al-Kindi, A. Y. (2009). Ultrastructural features and elemental distribution in eggshell during pre and post hatching periods in the green turtle, *Chelonia mydas* at Ras Al-Hadd, Oman. *Tissue and Cell,* 41, 214-221.

[87] Farre, B.; Brunelle, A.; Laprevote, O.; Cuif, J.-P.; Williams, C. T.; Dauphin, Y. (2011). Shell layers of the black-lip pearl oyster *Pinctada margaritifera*: Matching microstructure and composition. *Comparative Biochemistry and Physiology B*, 159, 131-139.

[88] Morgan, A. J.; Kille, P.; Bennett, A.; O'Reilly, M.; Fisher, P., Charnock, J. M. (2013). Pb and Zn imaging and *in situ* speciation at the geogenic/biogenic interface in sentinel earthworms using electron microprobe and synchrotron micro-focus X-ray spectrometry. *Environmental Pollution*, 173, 68-74.

[89] Molnar. L.; Engelmann, P.; Somogyi, I.; Macsik, L. L., Pollak, E. (2012). Cold-stress induced formation of calcium and phosphorus rich chloragocyte granules (chloragosomes) in the earthworm *Eisenia fetida*. *Comparative Biochemistry and Physiology A,* 163, 199-209.

[90] Aranyosiova, M.; Michalka, M.; Kopani, M.; Rychly, B; Jakubovsky, J.; Velic, D. (2008). Microscopy and chemical imaging of Behcet brain tissue. *Applied Surface Science*, 255, 1584-1587.

[91] Mauritz, J. A.; Seear, R.; Esposito, A.; Kaminski C. F.; Skepper, J. N.; Warley, A.; Lew, T.; Tiffert, T. (2011). X-ray microanalysis investigation of the changes in Na,K, and hemoglobin concentration in *Plasmodium falciparum*-infected red blood cells. *Biophysical Journal,* 100, 1438-1445.

[92] Elder, F. R.; Gurewitsch, A. M.; Langmuir, R. V.; Pollock, H. C. (1947). Radiation from electrons in a synchrotron. *Physical Review*, 71, 829-830.

[93] Stampanoni, M.; Groso, A.; Isenegger, A.; Mikuljan, G.; Chen, Q.; Bertrand, A.; Henein, S.; Betemps, R.; Frommherz, U.; Bohler, P.; Meister, D.; Lange, M.; Abela, R. (2006). Trends in synchrotron-based tomographic imaging: the SLS experience. *Proceedings of SPIE,* 6318, 1-14.

[94] Srajer, V.; Teng, T.; Ursby, T.; Pradervand, C.; Ren, Z.; Adachi, S.; Schildkamp, W.; Bourgeois, D.; Wulff, M.; Moffat, K. (1996). Photolysis of the carbon monoxide complex of myoglobin: nanosecond time-resolved crystallography. *Science*, 274, 1726-1729.

[95] Verbeni, R.; Pylkkanen, T.; Huotari, S.; Simonelli, L.; Vanko, G.; Martel, K.; Henriquet, C.; Monaco, G. (2009). Multiple-element spectrometer for non-resonant inelastic X-ray spectroscopy of electronic excitations. *Journal of Synchrotron Radiation*, 16, 469-476.

[96] Rao, D. V.; Swapna, M.; Cesareo, R.; Brunetti, A.; Akatsuka, T.; Yuasa, T.; Takeda, T.; Tromba, G.; Gigante, G. E. (2009). Investigation of the

distribution of elements in snail shell with the use of synchrotron-based, micro-beam X-ray fluorescence spectrometry. *Journal of Trace Elements in Medicine and Biology,* 23, 251-257.

[97] Bluet, P.; Lemelle, L.; Tucoulou, R.; Gergaud, P.; Delette, G.; Cloetens, P.; Susini, J.; Simionovici, A. (2013). 3D Chemical imaging based on a third-generation synchrotron source. *Trends in Analytical Chemistry*, 29, 518-527.

[98] Ortega, R.; Bohic, S.; Tucoulou, R.; Somogyi, A.; Deves, G. (2004). Microchemical element imaging of yeast and human cells using synchrotron X-ray microprobe with Kirkpatrick-Baez optics. *Analytical Chemistry,* 76, 309-314.

[99] Snigirev, A.; Kohn, V.; Snigireva, I.; Lengeler, B. (1996). A compound refractive lens for focusing high-energy X-rays. *Nature*, 384, 49-51.

[100] Di Fabrizio, E.; Romanato, F.; Gentili, M.; Cabrini, S.; Kaulich, B.; Susini, J.; Barrett, R. (1999). High-efficiency multilevel zone plates for keV X-rays. *Nature*, 401, 895-898.

[101] Lopes, R. T.; Lima, I.; Pereira, G. R.; Perez, C. A. (2011). Synchrotron radiation X-ray microfluorescence techniques and biological applications. *Pranama – Journal of Physics*, 76, 271-279.

[102] Stern, E. A.; Sayers, D. E. (1975). Extended X-ray-absorption fine-structure technique. III. Determination of physical parameters. *Physical Review B,* 11, 4836-4846.

[103] Benfatto, M.; D'Angelo, P.; Longa, S. D.; Pavel, N. V. (2002). Evidence of distorted fivefold coordination of the Cu^{2+} aqua ion from an X-ray-absorption spectroscopy quantitative analysis. *Phyiscal Review* B, 65, 174205.

[104] Zhao, Y.; Parsons, J. G.; Peralta-Videa, J. R.; Lopez-Moreno, M. L.; Gardea-Torresday, J. L. (2009). Use of synchrotron- and plasma-based spectroscopic techniques to determine the uptake and biotransformation of chromium(III) and chromium(VI) by *Parkinsonia aculeate. Metallomics*, 1, 330-338.

[105] Merdy, P.; Guillon, E.; Dumonceau, J.; Aplincourt, M. (2002). Spectroscopic study of copper(II)-wheat straw cell wall residue surface complexes. *Environmental Science and Technology*, 36, 1728-1733.

[106] Nuyts, G.; Cagno, S.; Hellemans, K.; Veronesi, G.; Cotte, M.; Jassens, K. (2013). Study of the early stage Mn intrusion in corroded glass by means of combined SR FTIR/μXRF imaging and XANES spectroscopy. *Procedia Chemistry*, 8, 239-247.

[107] Oyanagi, H.; Matsushita, T.; Tanoue, H.; Ishiguro, T.; Kohra, K. (1985). Fluorescence-detected X-ray absorption spectroscopy applied to structural characterization of very thin films; ion-beam induced modification of thin Ni layers on Si(100). *Japanese Journal of Applied Physics,* 24, 610-619.

[108] Goldberg, E. L.; Zolotarev, K. B.; Maksimovskaya, V. V.; Kondratyev, V. I.; Ovchinnikov, D. V.; Naurzbaev, M. M. (2007). Correlations and fixation of some elements in tree rings. *Nuclear Instruments and Methods in Physics Research A*, 575, 196-198.

[109] Mihucz, V. G.; Silversmit, G.; Szaloki, I.; de Samber, B.; Schoonjans, T.; Tatar, E.; Vincze, L.; Virag, I.; Yao, J.; Zaray, G. (2010). Removal of some elements from washed and cooked rice studied by inductively coupled plasma mass spectrometry and synchrotron based confocal micro-X-ray fluorescence. *Food Chemistry*, 121, 290-297.

[110] Guerra, C. S.; Gerlach, R. F.; Pinto, N. G. V.; Cardoso, S. C.; Moreira, S.; Almeida, A. P.; Peixoto, I. T. A.; Meloni, C. H.; Mota, C. L.; Oliveira, L. F.; Braz, D.; Barroso, R. C. (2010). X-ray fluorescence with synchrotron radiation to elemental analysis of lead and calcium content of primary teeth. *Applied Radiation and Isotopes*, 68, 71-75.

[111] Farquharson, M. J.; Geraki, K.; Falkenberg, G.; Leek, R.; Harris, A. (2007). The localisation and micro-mapping of copper and other trace elements in breast tumours using a synchrotron micro-XRF system. *Applied Radiation and Isotopes*, 65, 183-188.

[112] Al-Ebraheem, A.; Mersov, A.; Gurusamy, K.; Farquharson, M. J. (2010). Distribution of Ca, Fe, Cu and Zn in primary colorectal cancer and secondary colorectal liver metastases. *Nuclear Instruments and Methods in Physics Research A*, 619, 338-343.

[113] Banas, A.; Banas, K.; Falkenberg, G.; Dyduch. G.; Pawlicki, B.; Kwiatek, W. M. (2008). Using micro-synchrotron radiation induced X-ray emission distribution maps to determine correlation between elements in prostate tissue. *Spectrochimica Acta B*, 63, 957-961.

[114] Podgorczyk, M.; Kwiatek, W. M.; Grolimund, D.; Borca, C. (2009). Technical aspects of Zn microanalysis of human prostate cancer tissues and cells. *Radiation Physics and Chemistry*, 78, S53-S57.

[115] Ortega, R.; Bresson, C.; Fraysse, A.; Sandre, C.; Deves, G.; Gombert, C.; Tabarant, M.; Bleuet, P.; Seznec, H.; Simionovici, A.; Moretto, P.; Moulin, C. (2009). Cobalt distribution in keratinocyte cells indicates nuclear and perinuclear accumulation and interaction with magnesium and zinc homeostasis. *Toxicology Letters*, 188, 26-32.

[116] Gunter, K. K.; Aschner, M.; Miller, L. M.; Eliseev, R.; Salter, J.; Anderson, K.; Gunter, T. E. (2006). Determining the oxidation states of manganese in NT2 cells and cultured astrocytes. *Neurobiology of Aging*, 27, 18161826.

[117] Montarges-Pelletier, E.; Chardot, V.; Echevarria, G.; Michot, L. J.; Bauer, A.; Morel, J.-L. (2008). Identification of nickel chelators in three hyperaccumulating plants: An X-ray spectroscopic study. *Phytochemistry*, 69, 1695-1709.

[118] Shi, J.; Wu, B.; Yuan, X.; YY, C.; Chen, X.; Chen, Y.; Hu, T. (2008). An X-ray absorption spectroscopy investigation of speciation and biotransformation of copper in *Elsholtzia splendens. Plant Soil*, 302, 163-174.

[119] Lei, M.; Wan, X.-M.; Huang, Z.-C.; Chen, T.-B.; Li, X.-W.; Liu, Y.-R. (2012). First evidence on different transportation modes of arsenic and phosphorus in arsenic hyperaccumulator *Pteris vittata. Environmental Pollution,* 161, 1-7.

[120] Zhou, X.; Wang, L.; Sun, X.; Yang, X.; Chen, C.; Wang, Q.; Yang, X. (2011). Cinnabar is not converted into methylmercury by human intestinal bacteria. *Journal of Ethnopharmacology*, 135, 110-115.

[121] Baek, M.; Chung, H.-E.; Yu, J.; Lee, J.-A; Kim, T.-H.; Oh, J.-M.; Lee, W.-J.; Paek, S.-M.; Lee, J. K.; Jeong, J.; Choy, J.-H.; Choi, S.-J. (2012). Pharmacokinetics, tissue distribution, and excretion of zinc oxide nanoparticles. *International Journal of Nanomedicine*, 7, 3081-3097.

[122] Polgari, Z.; Meirer, F.; Sasamori, S.; Ingerle, D.; Pepponi, G.; Streli, C.; Rickers, K.; Reti, A.; Budai, B.; Szoboszlai, N.; Zaray, G. (2011). Iron speciation in human cancer cells by K-edge total reflection X-ray fluorescence – X-ray absorption near edge structure analysis. *Spectrochimica Acta* B, 66, 274-279.

[123] Strub, E.; Plarre, R.; Radtke, M.; Reinholz, U.; Riesemeier, H.; Schoknecht, U.; Urban, K.; Jungel, P. (2008). Determination of Cr(VI) in wood specimen: A XANES study at the Cr K edge. *Nuclear Instruments and Methods in Physics Research B*, 266, 2405-2407.

[124] Smith, E.; Kempson, I.; Juhasz, A. L.; Weber, J.; Skinner, W. M.; Grafe, M. (2009). Localization and speciation of arsenic and trace elements in rice tissues. *Chemosphere*, 76, 529-535.

[125] Cusack, M.; Dauphin, J.; Cuif, J.-P.; Salome, M.; Freer, A.; Yin, H. (2008). Micro-XANES mapping of sulphur and its association with magnesium and phosphorus in the shell of the brachiopod, *Terebratulina retusa. Chemical Geology*, 253, 172-179.

[126] Helmcke, K. J.; Syversen, T.; Miller, D. M.; Aschner, M. (2009). Characterization of the effects of methylmercury on *Caenorhabditis elegans*. *Toxicology and Applied Pharmacology*, 240, 265-272.

[127] Helmcke, K. J.; Aschner, M. (2010). Hormetic effect of methylmercury on *Caenorhabditis elegans*. *Toxicology and Applied Pharmacology*, 248, 156-164.

[128] Caito, S. W.; Zhang, Y.; Aschner, M. (2013). Involvement of AAT transporters in methylmercury toxicity in *Caenorhabditis elegans*. *Biochemical and Biophysical Research Communications*, 435, 546-550.

[129] Lin, Y.-T.; Hoang, H.; Hsieh, S. I.; Rangel, N.; Foster, A. L.; Sampayo, J. N.; Lithgow, G. J.; Srinivasan, C. (2006). Manganous ion supplementation accelerates wild type development, enhances stress resistance, and rescues the life span of a short – lived *Caenorhabditis elegans* mutant. *Free Radical Biology and Medicine*, 40, 1185-1193.

[130] Li, W.-H.; Hsu, F.-L.; Liu, J.-T.; Liao, V. H.-C. (49). The ameliorative and toxic effects of selenite on *Caenorhabditis elegans*. *Food and Chemical Toxicology*, 49, 812-819.

[131] Wan, L.; Nie, G.; Zhang, J.; Luo, Y.; Zhang, P.; Zhang, Z.; Zhao, B. (2011). β-Amyloid peptide increases levels of iron content and oxidative stress in human cell and *Caenorhabditis elegans* models of Alzheimer disease. *Free Radical Biology and Medicine*, 50, 122-129.

[132] Kim, J.; Shirawasa, T.; Miyamoto, Y. (2010). The effect of TAT conjugated platinum nanoparticles on lifespan in a nematode *Caenorhabditis elegans* model. *Biomaterials*, 31, 5849-5854.

[133] Zhao, Y.; Wu, Q.; Tang, M.; Wang, D. (2013). The *in vivo* underlying mechanism for recovery response formation in nano-titanium dioxide exposed *Caenorhabditis elegans* after transfer to the normal condition. *Nanomedicine: Nanotechnology, Biology, and Medicine, In Press*.

[134] Yeo, M.-K.; Nam, D.-H. (2013). Influence of different types of nanomaterials on their bioaccumulation in a paddy microcosm: A comparison of TiO_2 nanoparticles and nanotubes. *Environmental Pollution,* 178, 166-172.

[135] Davydkova, I. L.; Fadeeva, N. P.; Kovekovdova, L. T.; Fadeev, V. I. (2005). Heavy metal contents in tissues of dominant species of the benthos and in bottom sediments of Zolotoi Rog Bay, Sea of Japan. *Russian Journal of Marine Biology*, 31, 176-180.

[136] Zeitoun-Ghandour, S.; Charnock, J. M.; Hodson, M. E.; Leszczyszyn, O. I.; Blindauer, C. A.; Stürzenbaum, S. R. (2010). The two *Caenorhabditis*

elegans metallothioenins (CeMT-1 and CeMT-2) discriminate between essential zinc and toxic cadmium. *FEBS Journal*, 277, 2531-2542.

[137] Savoly, Z.; Nagy, P.; Havancsak, K.; Zaray, G. (2012). Microanalytical investigation of nematodes. *Microchemical Journal*, 105, 83-87.

[138] Howell, R. (1982). Levels of heavy metal pollutants in two species of marine nematodes. *Marine Pollution Bulletin*, 11, 396-398.

[139] Howell, R. (1983). Heavy metals in marine nematodes: uptake, tissue distribution and loss of copper and zinc. *Marine Pollution Bulletin*, 14, 263-268.

[140] Howell, R. (1982). The secretion of mucus by marine nematodes (*Enoplus* spp.): a possible mechanism influencing the uptake and loss of heavy metal pollutants. *Nematologica*, 28, 110-114.

[141] Jackson, B. P.; Williams, P. L.; Lanzirotti, A.; Bertsch, P. M. (2005). Evidence for biogenic pyromorphite formation by the nematode *Caenorhabditis elegans*. *Environmental Science and Technology*, 39, 5620-5625.

[142] Gao, Y.; Liu, N.; Chen, C.; Luo, Y.; Li, Y.; Zhang, Z.; Zhao, Y.; Zhao, B.; Iida, A.; Chai, Z. (2008). Mapping technique for biodistribution of elements in a model organism, *Caenorhabditis elegans*, after exposure to copper nanoparticles with microbeam synchrotron radiation X-ray fluorescence. *Journal of Analytical Atomic Spectrometry*, 23, 1121-1124.

[143] Savoly, Z.; Pepponi, G.; Nagy, P. I.; Streli, C.; Buzanich, G.; Chinea-Cano, E.; Zaray, G. (2013). Investigation of distribution and oxidation state of copper in soil-inhabiting nematodes by means of synchrotron radiation. *X-ray Spectrometry*, 42, 321-329.

[144] Qu, Y.; Li, W.; Zhou, Y.; Liu, X.; Zhang. L.; Wang, L.; Li, Y.-F.; Iida, A.; Tang, Z.; Zhao, Y.; Chai, Z.; Chen, C. (2011). Full assessment of fate and physiological behavior of quantum dots utilizing *Caenorhabditis elegans* as a model organism. *Nano Letters*, 11, 3174-3183.

[145] McColl, G.; James, S. A.; Mayo, S.; Howard, D. L.; Ryan, C. G.; Kirkham, R.; Moorhead, G. F.; Paterson, D.; de Jonge, M. D.; Bush, A. I. (2012). *Caenorhabditis elegans* maintains highly compartmentalized cellular distribution of metals and steep concentration gradients of manganese. *PLoS ONE*, 7, e32685.

[146] James, S. A.; de Jonge, M. D.; Howard, D. L.; Bush, A. I.; Paterson, D.; McColl, G. (2013). Direct *in vivo* imaging of essential bioinorganics in *Caenorhabditis elegans*. *Metallomics*, 5, 627-635.

[147] Nicholas, W. L.; Goodchild, D. J.; Stewart, A. (1987). The mineral composition of intracellular inclusions in nematodes from thiobiotic mangrove mud-flats. *Nematologica*, 33, 167-179.

[148] Williams, M. S. R.; Seraphin, S. (1998). Heavy metal biomineralization in free-living nematodes, *Panagrolaimus* spp. *Materials Science and Engineering C,* 6, 47-51.

[149] Savoly, Z.; Nagy, P.; Varga, G.; Havancsak, K.; Hracs, K.; Zaray, G. (2013). A novel method for investigation of uptake and distribution of polluting microelements and nanoparticles in soil-inhabiting nematodes. *Microchemical Journal*, 110, 558-567.

[150] Bongers, T.; Bongers, M. (1998). Functional diversity of nematodes. *Applied Soil Ecology*, 10, 239-251.

[151] Leapman, R. D.; Kocsis, E.; Zhang, G.; Talbot, T. L.; Laquerriere, P. (2004). Three-dimensional distributions of elements in biological samples by energy-filtered electron tomography. *Ultramicroscopy*, 100, 115-125.

[152] Cranfield, C. G.; Dawe, A.; Karloukovski, V.; Dunin-Borkowski, R. E.; de Pomerai, D.; Dobson, J. (2004). Biogenic magnetite in the nematode *Caenorhabditis elegans*. *The Royal Society Biology Letters*, 271, S436-S439.

In: Nematodes
Editor: Lee M. Davis
ISBN: 978-1-62948-764-9

Chapter 5

PLANTS AS NATURAL SOURCES OF NEMATICIDES

Marek Renčo[1*], Nicola Sasanelli[2] and Lara Maistrello[3]
[1]Parasitological Institute of Slovak Academy of Sciences, Košice, Slovak Republic
[2]Institute for Sustainable Plant Protection, C.N.R., Bari, Italy
[3]Department of Life Sciences, University of Modena and Reggio Emilia, Area S. Lazzaro, Pad. Besta, Reggio Emilia, Italy

ABSTRACT

Plant parasitic nematodes can cause serious damages to numerous economically important agricultural crops worldwide. Control of these pests is based on pesticides and on the use of resistant cultivars. Although chemicals are effective in the plant parasitic nematode control, the recent European legislation has deeply revised and restricted their use on agricultural crops focusing the attention on environmental safety and human and animal health. Moreover, resistant cultivars are available only for a limited number of crops and plant parasitic nematode species. During the last decade research on low environmental impact alternatives to nematicides has received a strong impulse and a wide range of options including agronomic strategies, physical methods, bio-pesticides, and biocidal plants haveas been considered. Focusing on plants derived

* Corresponding author: E-mail: renco@saske.sk.

natural plant products known as botanic pesticides can represent useful tools to minimize soil nematode population density under the tolerance limits of the different crops to specific phytoparasitic nematodes. These botanic pesticides are generally non-persistent in field conditions as they are readily transformed by light, oxygen and microorganisms into non-toxic products without residues in the environment. Many botanic pesticides derived from wild and cultivated plants have been shown to possess nematicidal properties against several plant-parasitic nematodes. The aim of this review is to report the effects of botanic pesticides derived from various wild, cultivated and medicinal plants, including essential oils, root and tissues extracts, on plant parasitic nematodes in the view of sustainable agriculture.

INTRODUCTION

Nematodes are a highly assorted group of invertebrates occupying many terrestrial and aquatic habitats. Among the soil inhabiting organisms the nematodes dominate over all others, both in number and species (Ingham, 1998). They are small multi-cellular animals with a rather simple organization and a permeable body wall cuticle, so that they are in direct contact with substances dissolved in the capillary soil water. Throughout the evolution, several groups of soil nematodes have developed in respect to their feeding habits, so that they can by distinguished in those that feed on bacteria, fungi, animals, algae or plants. The focus of this chapter is the sustainable management of those nematodes that depend on plants as their food source, the plant parasitic nematodes (also called phytonematodes), which are a major group of soil nematodes that includes species that can cause considerable damage to many important agricultural and forest crop plants. By means of the buccal needle, the "stylet", plant parasitic nematodes pierce the cuticle cells of plant tissues (eg. young roots) and feed on the lymph, simultaneously injecting enzymes that are toxic to the plant cells. Moreover, several species of plant parasitic nematodes of the genera *Longidorus, Trichodorus* or *Xiphinema* could transport in their saliva different kinds of plant viruses which multiply the harmfulness (Brown et al., 1994). Worldwide, there are numerous genera and species of plant parasitic nematodes. Among them, there are some species occurring in all continents, except very cold areas, that can attack hundreds of different plant species, such as the root-knot nematodes belonging to *Meloidogyne* genus. On the opposite, there are highly specialized species capable to attack only few botanically related species, as in the case of the cyst forming nematodes of the genera *Heterodera* and *Globodera*.

Over 4,100 species of plant-parasitic nematodes have been identified (Decraemer & Hunt, 2006) and species of economic importance can be grouped into relatively restricted specialized groups that either cause direct or indirect damage to their host. These nematodes attack roots and below-ground tissues of plants, although some species can infect plant foliage, stems and flowers (Sasanelli et al., 1997; Lamberti et al., 2001). When plant parasitic nematodes increase their abundance up to the threshold level, plant damage could be so high to cause yield losses ranging in millions dollars, depending on the crop. Nevertheless, data on their economic impact remain less than comprehensive, especially for crops produced in resource poor areas. Some estimates suggest they cause 80 billion dollars of damage worldwide each year (Handoo, 1998) so that control measures are necessary.

Various methods are commonly used to control plant-parasitic nematodes. These methods can be divided into three main types: biological, cultural and chemical control. The only type of strategy that has been successfully used for to control of plant parasitic nematodes in the past 50 years was the use of chemical nematicides that effectively kill nematodes in soil (Lambert & Bekal, 2002) and can be distinguished in soil fumigants (gas) and non-fumigant nematicides (liquid or solid formulations). Soil fumigants became popular because they drastically reduced nematode populations in the soil without requiring alternative host crops for rotation and were usually cost effective for most crops. However, some fumigant nematicides have been banned since year 1979 (DBCP) by the U.S. Environmental Protection Agency as environmental toxins, with the exception of 1,3 dichloropropene (Telone II), chloropicrin (tear gas), and dazomet (Basamid) (Sittig, 1985). The multipurpose soil fumigant methyl bromide also provided an excellent reduction of soil nematode populations, but its use was largely discontinued in the world after 2005. Non-fumigant nematicides such as fenamiphos and aldicarb are based upon the same kind of active ingredients of many insecticides (i.e. nerve poisons) and can be applied in liquid or granular formulations (Lambert & Bekal, 2002).

Also the European Legislation has deeply restricted and revised the use of pesticides on agricultural crops focusing the attention on environmental safety, human and animal health (Reg. CE 396/2005; 1095/2007; 33/2008; 299/2008) and, in particular the recent framework directive on the sustainable use of pesticides, 2009/128/EC fosters the development of plant protection and integrated pest management (IPM) in the EU, -stating that "*when pesticides are used, appropriate risk management measures should be established and low-risk pesticides as well as biological control measures should be considered in the first place"-*. It has been estimated that about 2.5 million tons of pesticides are used on crops each year and the worldwide damage caused by pesticides reaches $100 billion annually (Koul et al., 2008), of which a significant percentage is represented by nematicides. For example, according to Shoemaker & Been (1999) in the Netherlands they represent more than 60% of the total pesticides used in agriculture. There is therefore an urgent need to replace pesticides with alternative means of control that are less toxic and more environmentally friendly.

As a consequence, research of alternative methods for the sustainable management of plant pests, including parasitic nematodes, has received a strong impulse. Among them there are organic amendments, soil solarisation, resistant plant varieties, use of antagonists (eg. predators) or pathogens (fungi and bacteria). However, the most studied are natural plant products, known as botanical pesticides or phytochemicals. These are excellent candidates because they can be developed for use as nematicides themselves, or they can serve as model compounds for the development of chemically synthesised derivatives which enhanced activity or environmental friendliness (Chitwood, 2002). Use of botanicals is now emerging as one of the important means to be used in protection of crop produce and the environment from pesticides pollution, which is a global problem (Isman, 2006; Ntalli & Menkissoglu-Spiroudi, 2011; Ntalli & Caboni, 2012). On the other side the most, although not all the discovered phytochemicals are definitely safer to the environment and humans than the traditional chemical nematicides (Gommers, 1981; Graine & Ahmed, 1988). Therefore it is important to examine new plants or new varieties of plants for their efficacy in immobilizing, retarding development or killing nematodes. The present chapter reviews the use of plants as sources of compounds to be used in the sustainable management of plant parasitic nematodes, focusing especially on the species of major economic relevance worldwide, such as the root-knot nematodes of the genus *Meloidogyne* that have a wide range of hosts and cause serious problems to many annual and perennial crops.

Plant Secondary Metabolites

Higher plants are able of to synthesize and produce a variety of small organic molecules called secondary metabolites (Sarker et al., 2005). These metabolites contains products that may help in the growth and development of plants, however they are not essential for the plant to survive. Secondary metabolism facilitates the primary metabolism in plants (Kutchan & Dixon, 2005). Plants and their secondary metabolites are an important source for biopesticides and for the development of new pesticides. The recognition of the important role of these compounds has increased, particularly in terms of resistance to pests and diseases. The intensive use of synthetic pesticides and their environmental and toxicological risks have generated increased global interest in developing alternative sources of chemicals to be used in safe management of plant pests (Cavoski et al., 2011; 2012).

Secondary metabolites often play an important role in plant defence against herbivory and other interspecies defences (Stamp, 2003). These chemicals are extremely diverse and many thousands have been identified in several major classes. The most important are alkaloids, monoterpenoids, flavonoids, diterpenoids, and polyphenols. Among polyphenols, tannins protect several plants against herbivores (Feeny, 1976) and are toxic to a wide range of fungi, bacteria and yeasts (Scalbert, 1991). Chemical and physical properties of tannins are highly various according to the plant species (i.e. *Castanea sativa* Miller or *Schinopsis lorentzii* (Schldl.) Barkley & Meyer, etc.), the plant parts and the season in which they are produced (Waterman, 1999; Waghorn & McNabb, 2003). Tannins have been reported to possess antihelmintic activity for gastrointestinal nematodes in ruminants (Hoste et al., 2006) and their effect against plant parasitic nematodes is also promising in the view of sustainable agriculture.

Taylor and Murant (1966) found a reduction of *Longidorus elongatus* in soil treated with powdered tannin extracts from mimosa (*Acacia mollisima* L.) and quebracho (*Schinopsis lorentzii* L.); amendments rich in tannins proved to be effective against *Meloidogyne arenaria* (Mian and Rodríguez-Kábana, 1982), and our recent experimental findings show that chestnut (*Castanea sativa* L.) tannins aqueous solutions are effective in controlling the root-knot nematode *M. javanica,* the potato cyst nematode *Globodera rostochiensis* patotype Ro1, *G. pallida* patotype Pa2 and the carrot cyst nematode *Heterodera carotae* (Maistrello et al., 2010; 2013; Renčo et al., 2012, 2013). In the *in vitro* experiments on both nematode species *M. javanica* and *G. rostochiensis,* chestnut tannin solutions in geometrical scale concentrations from 0.32 to 20.48 g/l significantly greatly reduced nematode egg hatch in comparison to untreated control during the incubation period with addition of root leachate. After removal of the tannin solutions when nematode incubation continued only in tomato or potato root leachate, the emergence of juveniles remained significantly lower compared to the control until to the end of the experiment. Similar results were obtained in the *in vivo* glasshouse experiments in tomato/potato pots artificially infested with *M. javanica*/*G. rostochiensis,* when all tested doses of chestnut tannin (100, 250 and 450 g/m^2) significantly reduced the population and reproduction rate of both nematode species in comparison to infested untreated control.

The nematicidal effects of alkaloids, saponins, monoterpenoids, pentacyclic, triterpenoid, triglycerides, sesquiterpenes, steroids, diterpenes, flavonoid or glucosinolates have been also found (Chitwood, 2002). These compounds can be developed for use as nematicides themselves, or can serve as model compounds for the development of chemically synthesized derivatives with enhanced activity and reduced environmental impacts. For example, Wang et al. (2012) recorded that three alkaloids extracted from *Macleaya cordata* had strong nematicidal activity to *B. xylophilus, Caenorhabditis elegans* and *M. incognita*: they were sanguinarine at LC_{50} 28.5, 22.8 and 67.5 μg/ml respectively.; chelerythrine atLC_{50} 34.5, 40.3 and 61 μg/ml respectively.; and allocryptopine LC_{50}37.5, 38.9 and 76.6 μg/ml respectivelly. Similarly Thoden et al. (2007, 2009) found that 1,2-dehydropyrrolizidine alkaloids extracted from *Chromolaena odorata* invasive weeds, *Crotalaria* spp. or *Ageratum* spp. had nematicidal, ovicidal and repellent effects on different plant-parasitic and free-living nematodes including *M. incognita*. Begum et al. (2008) observed that the triterpens pomolic acid, lantanolic acid, and lantoic acid showed 100% mortality of *M. incognita* at 1 mg/ml concentration after 24 h, while camarin, lantacin, camarinin, and ursolic acid exhibited 100% mortality at this concentration after 48 h. These results are comparable to those obtained with the conventional nematicide furadan (100% mortality at 1 mg/ml concentration after 24 h). The Pphenolic monoterpene named thymol possessses nematicidal effects against *M. arenaria* at estimated concentration in soil of 85 ppm (LC_{50}) and against or *H. glycines* at 145 ppm (LC_{50}) (Soler-Serratosa et al., 1996). Secondary metabolites of rye (*Secale cereale*) such as cyclic hydroxamic acids also showed nematicidal activity against *Xiphinema americanum* and *M. incognita* (Zasada et al., 2005). Eggs of *M. incognita* were less sensitive to the hydroxamic acids than J2. Lazerri et al. (1993) observed, that while glucosinolates extracted from the hosts of *Heterodera schachtii* i.e. *Brassica napus, B. rapa, B. corinata, Lepidum sativum, Raphanus sativus* and *Sinapis alba* showed no nematicidal effect against *H. schachtii* this nematode, on the opiste their derivates, products of the enzymatic hydrolysis at pH 7.0 (essentially isothiocyanate), e.g. such as sinirgin, gluconapin, glucotropeolin, glucode-hydrorucin, and the whole glucosinolate group extracted from rapessed caused mortality of this nematode. Also saponins from *Medicago arborea* (tops), *M. arabica* (tops and roots) and *M. sativa* (tops and roots) induced 100% mortality of *Xiphinema index* at 500μg ml^{-1}, while prosapogenins resulted in toxicity, starting at 125μg ml^{-1} (Argentieri et al., 2012). Saponins from *M. sativa* were found to be nematotoxic also to the root-knot nematode *M. incognita* (100% mortality at 500μg ml^{-1}) and the potato cyst nematode *Globodera rostochiensis* (maximum 68-91% mortality at 250-1,000 μg ml^{-1}) as well as to *X. index* in the study of D′Addabbo et al. (2010). Similarly, triterpenoid saponins isolated from the roots of *Pulsatilla koreana* showed strong nematicidal activity against *M. incognita* after 48 h, with an LC50 value of 92.8 μg/ ml^{-1}, which suggests that triterpenoid saponins from *P. koreana* should be explored as potential natural nematicides for developing new agents to control root-knot nematode disease (Li et al., 2013) and the more promising are raddeanoside R13, hederoside C, and pulsatilla saponin D.

All these bioassays showed that secondary plant metabolites affect the behaviour of plant parasitic nematodes, acting as attractants or repellents, inducing paralysis, reducing hatch or even causing death. More investigations on the effects of secondary plant metabolites derived from different plant species and their parts on different plant parasitic nematode species as well as on beneficial soil free living nematodes need to be performed. A better understanding of the interaction between plant parasitic nematodes and plant secondary metabolites could contribute to increase plant defence against nematodes in resistant cultivars.

Herbal Powders as Plant Parasitic Nematodes Suppressant

Among plant parasitic nematodes, the root-knot nematodes of the genus *Meloidogyne*, mainly the species *M. incognita* (Kofoid *et* White) Chitw., *M. hapla* Chitwood, *M. javanica* (Treub) Chitwood and *M. arenaria* (Neal) Chitwood are considered the most economically important in agriculture. As an example, according to Jain et al. (1994), *M. incognita* is responsible for yield losses of up to 30% under field conditions. For this reason, different studies were targeted mainly on these nematodes and the nematicidal effect of hundred types of powders, prepared from different parts of plants was investigated. These powders were applied to the soil at different rates in the solid state or as aqueous solution when powders were diluted in water (Table 1).

Table 1. List of plant species and part of plant whose powder showed nematicidal activity against root-knot nematode species

Nematode	Plant species	Part for extraction	Effective concentration	References
Meloidogyne javanica	*Azadirachta indica*	seeds, leaves	30 ml/250 g soil	Moosavi, 2012
	Datura metel	seeds, leaves	30 ml/250 g soil	
	Olea europaea	leaves	30 ml/250 g soil	
	Carthamus glaucus	seeds	30 ml/250 g soil	
	Neria oleander	flowers	30 ml/250 g soil	
	Chrysanthemum coroniarum	flowers	30 ml/250 g soil	
	Hypericum androsaemum	whole plant	20 μg ml^{-1}	Al-Banna et al., 2003
	Anagyris foetida	leaves, fruits	20 μg ml^{-1}	
	Thea sinensis	leaves	20 μg ml^{-1}	
	Varthemia iphionoides	leaves	20 μg ml^{-1}	
	Euphorbia macroclada	leaves	20 μg ml^{-1}	
	Ferula harmonis	roots	20 μg ml^{-1}	
	Gundelia tournifortii	whole plant	20 μg ml^{-1}	
	Lepidium sativum	fruits	20 μg ml^{-1}	
	Pimpinella anisum	fruits	20 μg ml^{-1}	
	Inula viscosa	leaves	0.1% (w/w)	Oka et al., 2001
Meloidogyne incognita	*Nicotiana tabacum*	leaves	1.9 mg ml^{-1} (LC_{50})	Wiratno et al.,2009
	Syzygium aromaticum	bud	3.9 mg ml^{-1} (LC_{50})	
	Piper betle	leaves	3.0 mg ml^{-1} (LC_{50})	

Nematode	Plant species	Part for extraction	Effective concentration	References
	Acorus calamus	roots	11.3 mg ml^{-1} (LC_{50})	
	Azadirachta indica	leaves	2 g kg^{-1} soil	Agbenin et al., 2004
	Artemisia herba-alba	leaves	20 µg ml^{-1}	Al-Banna et al.,2003
	Euphorbia macroclada	leaves	20 µg ml^{-1}	
	Pimpinella anisum	fruits	20 µg ml^{-1}	
	Nepeta cataria	whole plant	50 ppm (LC_{50})	Pavaraj et al., 2012
	Couroupita quianensis	whole plant	80 ppm (LC_{50})	
	Pentanema indicum	whole plant	80 ppm (LC_{50})	
	Foeniculum vulgare	cloves	43 ppm (LC_{50})	Ibrahim et al., 2006
	Pinus pinea	leaves	44 ppm (LC_{50})	
	Origanum syriacum	leaves	50 ppm (LC_{50})	
	Mentha microcorphylla	leaves, flowers	65 ppm (LC_{50})	
	Eucalyptus spp	leaves	66 ppm (LC_{50})	
	Citrus sinensis	leaves	121 ppm (LC_{50})	

In addition to studies where root- knot nematodes were suppressed by powders of different plant species, in several works toxicity of plant powders on other plant parasitic nematodes was investigated. Kowalska and Smolinska (2001) found that volatile compounds (isothiocyanates) from seeds of *Brassica juncea, B.rassica napus* and *S.inapis alba* exhibit toxic activity against *Paratylenchus* spp., and populations of *Aphelenchoides composticola* Franklin infesting white button mushrooms were significantly reduced by leaf, flower or seed extracts of *Bougainvillea spectabilis, Calotropis procera, Cedrella toona, Jacarand acutifolia, Melia azadirach, Ricinus communis* and *Tagetes patula* (Grewal, 1989), with *M. azadirach* and *R. communis* seed extracts being highly toxic to this nematode and to *M. incognita* (Cavoski et al., 2012).

Also, *Tylenchulus semipenetrans* Cobb, *Ditylenchus dipsaci* (Kuhn) Filipjev and *Pratylenchus mediterraneus* Corbett were suppressed by leaf extracts of *Inula viscosa* (Oka et al., 2001).

Aqueous Plant Extracts with Nematicidal Properties

Besides the studies of potential nematicidal effects of plant powders, many investigations were addressed on the effects of fresh or methanol aqueous extracts of leaves or roots of various plants against several plant parasitic nematode species.

Bhattacharya et al., (2012) found that leaf extract of *Glycosmis pentaphylla* (Tamil) and bark of *Holarrbena antidysenterica* (Tellicherry) showed strong nematicidal properties against populations of *M. incognita* attacking tea. The percentage mortality of the nematode juveniles increased with the rise in concentration of the plant extracts and time of exposure moreover amendment of the soil with chopped leaves of *G. pentaphylla* reduced the reproduction of *M. incognita* and improved the growth of tea more than did bark extract of *H. antidysenterica*. Besides, the leaf extract of the medicinal plant *Glycosmis aculata* have been reported to possess nematicidal activity against *Radopholus similis* Cobb (Jasy & Koshy, 1992).

Hatching of *M. incognita* juveniles was significantly reduced also in root extracts of *Chromolaena odorata* (Siam weed), *Azadirachta indica* (Neem), *Ricinus communis* (Castor bean) and *Cymbopogon citratrus* (Lemon grass) (Adegbite & Adesiyan, 2005); *Tagetes erecta* (Marigold) (Sasanelli & D´Addabbo, 1993b), *Hyptis suaveolens* (Nitta) and *Ocimum gratisimmum* (Basil) (Olabiyi, 2008); water extract of seed, leaf and bark of *Tamarindus indica* (Tamarind), *Cassia siamea* (Kassod tree), *Isoberlinia doka* (Doka), *Dolnix regia* (Flamboyant) and *Cassia sieberiana* (African laburnum) (Bello et al., 2006); aqueous extracts of *Cannabis sativa* (Hemp) and *Zanthoxylum alatum* (Toothache Tree) (Mukhtar et al., 2013); aqueous extracts of leaves and roots of *Ruta graveolens* (common Rue) (Sasanelli & D´Addabbo, 1993a; 1993b); aqueous leaf extracts of *Murraya koenigii* (Curry leaf), *Jasminum sambac* (Jasmine), *Citrus aurantiifolia* (Sour orange), *Rauvolfia serpentina* (Patal garuda), *Zizyphus jujuba* (Ber), *Hibiscus rosa-sinensis* (China rose) and *Justicia gandurosa*(Padhi & Behera, 2000); seeds extracts of *Punica granatum, Thymus vulgaris* and *Artemisia absinthium* (Koryaem et al., 1993); extracts from *Calea urticifolia* leaves and roots, *Eugenia winzerlingii* leaves, and *Tephrosia cinerea* stems (Alejoet al.,2006),; *Dinbera retroflexa* (leaves),

Cucumis melo var. *agrestis* (fruits), *Eucalyptus microtheca* (leaves), *Acacia nilotica* (pods), and *Chenopodium album* (leaves) extracts (Elbadri et al., 2008b); cold aqueous extracts of whole plants, roots and stem of *Tagetes erecta* (Natarajan et al., 2006).

Aqueous extracts of leaves and roots of *Tagetes erecta* (marigold), *Cineraria maritima* (dusty miller) and *Ruta graveolens* (rue) showed good nematicidal activity against the sugar beet nematode *Heterodera schachtii* Schmidt, and on the root knot nematodes *M. arenaria, M. hapla, M. incognita* and *M. javanica* (Sasanelli & D´Addabbo, 1992, 1993b) . However, root extract of *C. maritima* had a greater nematicidal effect than its leaf extracts, probably because of the higher content of alkaloids (jacobine, jacodine and senecionine) in the roots (Grainge & Ahmed, 1988) than in the leaves. Volatile compounds of *Tagetes erecta* (myristic acid) showed higher nematicidal effect on *M. incognita* juveniles than its dodecanonic acid (Ray et al., 2000). However, leaf and root extracts of *Tagetes erecta* and *T.signata* showed no nematicidal effects when tested on the hatching of an Italian population of the potato cyst nematode *G. rostochiensis* (Sasanelli & Di Vito, 1991).

The ethanolic root extracts of *Artemisia vulgaris*showed inhibition of egg hatch (50% inhibition by 2.35 mg/ml) and caused second-stage juvenile mortality (50% lethality at 12 hours' exposure to 55.67 mg/ ml^{-1}) to *Meloidogyne megadora* Whitehad (Costa et al., 2003). Root extracts of *Acorus gramineus, Asiasarum sieboldi, Illicium verum* and *Kaempferia galanga* showed nematicidal activity against male, female and juveniles ofthe most dangerous pine wood nematode, *Bursaphelenchus xylophilus* (Steiner *et* Buhrer) Nickle (Choi et al., 2006),showing a potential in the management of this pest.

Among extracts from other plant species showing nematicidal effects on plant parasitic nematodes there are leaves and stems extracts of bean (*Phaseolus vulgaris* L.) and leaves of tobacco (*Nicotiana tabacum* L.) that showed toxic effects on *Tylenchorhynchus dubius* (Bütschli) Filipjev and *Hoplolaimus* spp. (Miller et al., 1973); neem (*Azadirachta indica*) leaf extracts were effective against *Pratylenchus brachyurus* (Godfrey) Filipjev *et* Shuurmans Stekhoven in maize field (Egunjobi & Afolami, 1976), and seed extracts from *Punica granatum, Thymus vulgaris* and *Artemisia absinthium* proved toxic to *Helicotylenchus dihystera* (Cobb) Sher (Korayem et al., 1993). Nematicidal effect of aqueous extracts of leaves of *Ruta graveolens* and *Thymus vulgaris* were evident also against *Xiphinema index* Thorne et Allen, *M. javanica* and *M. incognita*, respectively (Sasanelli, 1992; Sasanelli & D'Addabbo, 1993a; Sasanelli et al., 2007; 2009).

Essential Oils and their Nematicidal Activity

Essential oils (EOs) are defined as any volatile oil(s) that have strong aromatic components and that give distinctive odour, flavour or scent to a plant (Koul et al., 2008). These are the by-products of plant metabolism and are commonly included among volatile plant secondary metabolites. The aromatic characteristics of essential oils provide various functions for the plants including (i) attracting or repelling insects, (ii) protecting themselves from heat or cold; and (iii) utilizing chemical constituents in the oil as defence materials (Nerio et al., 2010). Many essential oils have been used as food additives, flavourings, and components of cosmetics, soaps, perfumes, plastics, and as resins (Koul et al., 2008). Different EOs have been found to possess antimicrobial and insecticidal activity (Oka et al.,2001), moreover many EOs and their components have been evaluated for their nematicidal properties (Table 2).

Table 2. Confirmed nematicidal effects of some plant essential oils against several plant parasitic nematodes

Nematode	Essential oil	Concentrations	References
Bursaphelenchus xylophilus	*Chamaespartium tridentatum*	1.981 mg ml^{-1} (LC_{100})	Barbosa et al., 2010
	Origanum vulgare	1.984 mg ml^{-1} (LC_{100})	
	Satureja montana	0.585 mg ml^{-1} (LC_{100})	
	Thymbra capitata	0.985 mg ml^{-1} (LC_{100})	
	Thymus caespititius	0.874 mg ml^{-1} (LC_{100})	
	Boswellia carterii	0.21 mg ml^{-1} (LC_{50})	Choi et al., 2007a
	Paeonia moutan	0.26 mg ml^{-1} (LC_{50})	
	Perilla frutescens	0.41 mg ml^{-1} (LC_{50})	
	Schizonepeta tenuifolia	0.41 mg ml^{-1} (LC_{50})	
	Brassica integrefolia	500 ppm (LC_{100})	Elbadri et al., 2008a
	Pelargonium inquinans	500 ppm (LC_{100})	
	Thymus vulgaris	500 ppm (LC_{100})	

Nematode	Essential oil	Concentrations	References
	Litsea cubeba	0.504 mg ml^{-1} (LC_{50})	Park et al., 2007
	Pimenta dioica	0.609 mg ml^{-1} (LC_{50})	
	Trachyspermum ammi	0.431 mg ml^{-1} (LC_{50})	
	Coriandrum sativum	-	Kim et al., 2008
	Liquidambar orientalis	-	
	Valeriana wallichii	-	
	Gaultheria fragrantissima	5 mg ml^{-1} (LC_{100})	Kim et al., 2011
	Zanthoxylum alatum	5 mg ml^{-1} (LC_{100})	
	Allium cepa	12.1-17.6 μg ml^{-1} (LC_{50})	Choi et al., 2007b
	Allium sativum	-	Park et al., 2005
	Cinnamonum verum	-	
Meloidogyne javanica	*Carum carvi*	100-200 mg kg^{-1} soil	Oka et al., 2000
	Coridothymus capitatus	100-200 mg kg^{-1} soil	
	Foeniculum vulgare	100-200 mg kg^{-1} soil	
	Mentha rotundifolia	100-200 mg kg^{-1} soil	
	Mentha spicata	100-200 mg kg^{-1} soil	
	Origanum syriacum	100-200 mg kg^{-1} soil	
	Origanum vulgare	100-200 mg kg^{-1} soil	
	Cymbopogon martini	4,000 μg ml^{-1} (LC_{95})	Sangwan et al.,1985
	Cymbopogon flexuosus	4,000 μg ml^{-1} (LC_{95})	
	Cymbopogon winterianus	4,000 μg ml^{-1} (LC_{55})	
	Cymbopogon martini	-	
	Eucalyptus citriodora	-	
	Eucalyptus hybrida	-	
	Mentha arvensis	-	
	Mentha piperita	-	

Table 2. (Continued)

Nematode	Essential oil	Concentrations	References
	Mentha spicata		Pandey et al. 2000
	Ocimum basilicum	-	
	Pelargonium graveolens	-	
	Mentha arvensis	0.291 mg ml^{-1} (LC_{50})	Andrés et al., 2012
	Mentha rotundifolia	0.204 mg ml^{-1} (LC_{50})	
	Mentha spicata	0.293 mg ml^{-1} (LC_{50})	
	Satureja montana	0.041 mg ml^{-1} (LC_{50})	
	Thymus mastichina	0.300 mg ml^{-1} (LC_{50})	
	Thymus vulgaris	0.224 mg ml^{-1} (LC_{50})	
	Thymus zygis	0.226 mg ml^{-1} (LC_{50})	
Meloidogyne incognita	*Pelargonium graveolens*	2,000 µg l^{-1}(LC_{100})	Leela et al., 1992
	Chenopodium ambrosioides	49.6 µg ml^{-1} (LC_{50})	Bai et al., 2011
	Agastache rugosa	47.3 µg ml^{-1} (LC_{50})	Li et al., 2013
Anguina tritici	*Cymbopogon martini*	4,000 µg ml^{-1} (LC_{95})	Sangwan et al., 1985
	Cymbopogon flexuosus	4,000 µg ml^{-1} (LC_{95})	
	Cymbopogon winterianus	4,000 µg ml^{-1} (LC_{55})	
Tylenchuslus semipenetrans	*Cymbopogon flexuosus*	1,000 µg ml^{-1} (LC_{90})	Sangwan et al., 1985
	Cymbopogon martini	1,000 µg ml^{-1} (LC_{100})	
	Cymbopogon winterianus	1,000 µg ml^{-1} (LC_{85})	
Heterodera avanae	*Cymbopogon flexuosus*	4,000 µg ml^{-1} (LC_{90})	Sangwan et al., 1985
	Cymbopogon martini	4,000 µg ml^{-1} (LC_{100})	
	Cymbopogon winterianus	4,000 µg ml^{-1} (LC_{65})	
	Hyssopus cuspidatus	338.70 µg ml^{-1} (LC_{50})	Li et al., 2012
Ditylenchus dipsaci	*Eugenia caryophyllata*	5,000 ppm (LC_{100})	Zouhar et al., 2009
	Origanum compactum	3,000-7,500 ppm (LC_{100})	
	Origanum vulgare	3,000-5,000 ppm (LC_{100})	
	Thymus matschiana	7,500 ppm (LC_{100})	
	Thymus vulgaris	1,500-7,500 ppm (LC_{100})	

Recently, Andrés et al. (2012) reviewed literature where the EOs were tested to their nematicidal effect. They stated that a large number of EOs extracted from different botanical families has been analyzed *in vitro* for nematicidal activity mainly against pine wood nematode *B. xylophilus* and *Meloidogyne* spp. The majority of studies evaluating the nematicidal activity of EOs against *B. xylophilus* have been carried out in Korea, where the pine wilt disease caused by the this nematode is the most serious problem in forests of the southern parts of the country (Andrés et al.,2012).

The mode of action of EOs against nematodes is still not fully understood (Ibrahim et al., 2006). In insects, several EOs inhibit acetylcholinesterase activity (Ryan & Byrne, 1988). Some common components of essential oils such as carvacrol, *t*-anethole, thymol, aliphatic aldehydes, phenols and monocyclic terpenes have been found to be insecticidal (Chen etal., 1997; Isman, 1999, Traboulsi et al., 2002; Sampson et al., 2005; Caboni et al., 2013), however carvacrol and thymol together with geraniol showed also high nematicidal activity against the plant parasitic nematodes *M. javanica* (Oka et al., 2000; Andrés et al.,2012), *M. incognita* (Al-Banna et al.,2003; Ibrahim et al., 2006) or *B. xylophilus* (Park et al., 2007). Sangwan et al. (1985) stated, that nematicidal activity of essential oils from palmarosa, lemon grass and citronella Java are thought to be caused by their characteristic mixture of monoterpenes. Lemon grass oil was effective against all four nematode tested (*M. javanica, H. avanae, T. semipenetrans, A tritici*) but citral, its principal constituent was effective only to *A tritici, H. avanae* and *T. semipenetrans*. The alcoholic constituents, geraniol and citronellol were effective against *M. javanica*, whereas the aldehydic constituent's citral and citronellal were much less effective.

Also othercomponents of EOs have shown nematicidal properties, such as methone and linalool (Oka et al., 2000, Ibrahim et al., 2006, Barbosa et al., 2010) against *M. javanica, M. incognita* or *B. xylophilus*; geranial, isoeugenol, methyl isoeugenol, eugenol, methyl eugenol and neral (Park et al.,2007; Li et al., 2012) against *B. xylophilus* or *Heterodera avanae*; benzaldehyde, trans-cinnamyl alcohol, cis-asarone, octanal, nonanal, decanal, trans-2-decenal, undecanal, dodecanal and decanol (Kim et al., 2008) or methyl and ethyl salicylates (Kim et al., 2011) to *B. xylophilus*; (*Z*)-ascaridole, ρ-Cymene and isoascaridole (Bai et al., 2011) or 2-thiophenecarboxaaldehyde and methylisothiocyanate (Caboni et al., 2012) against *M. incognita*.

In conclusion, essential oils and their components appear to be useful as natural nematicides against plant parasitic nematode species, mainly the root-knot nematodes of the genus *Meloidogyne* and the pine wood nematode *B.*

Xylophilus, which can attack pine trees also in protected areas where other methods of controls are not allowed. In order to preoceed with the practical use of the EOs and their components as novel nematicides, further study isstudies are necessary. Determination of the most appropriate rates and concentrations of the different essential oils and their feasibility for use in field or enclosed conditions are needed (Zouhar et al., 2009). Besides, other studies should focus on the systemic action, phytotoxicity and formulation to improve nematicidal potency and stability and to reduce the costs.

Green Manure versus Plant Extracts

Green manure, that is the practice to plowing under specific crops while they are green or soon after they flower, positively influences the biological soil activity because of decomposing plant material is a good source of nutrients and energy for the soil microorganisms. But it is also one way to add to the soil nematicidal substances, which are released during the decomposition of green plants after their ploughing. The nematicidal activity effect depends on the plants used for green manure and varies according to the parasitic nematode to manage.

Plants from the Brassicaceae family are the most commonly reported to suppress soil-borne pests and pathogens, including plant-parasitic nematodes, when used as soil amendments (Valdez et al.,2011). These suppressive effects can be obtained by increasing the presence of beneficial soil organisms (Ploeg, 2008) or through the effect known as 'biofumigation' (Kirkegaard et al., 1993). The biofumigant effect is obtained when glucosinolates and their isothiocyanates hydrolysed derivates (present in all members of the Brassicaceae) are released during the breakdown and soil incorporation of the plants (Valdez et al., 2011). Different species and varieties of brassicas can vary considerably in types and quantities of glucosinolates (Rosa et al., 1997), as well as quantities and types of isothiocyanates formed during the breakdown of the plants (Ploeg, 2008) and the nematicidal effects can be different according also to the nematode species.

As found by Kowalska and Smolinska (2001) *B.rassica napus* and *B.rassica juncea* volatile compounds increased population of *Paratylenchus* nematodes, whereas green manure with *S.inapis alba* decreased their population level. Valdes et al. (2011) showed that *B. napus, Raphanus sativus* and *S.inapis alba* plant extracts were not nematicidal on potato cyst nematode *G. rostochiensis*. These differences, with *S. alba* being nematicidal against

Paratylenchus but not against *G. rostochiensis* nematode can be attributed to the different biology of these nematodes as well as to the experimental design. Life cycle of *Paratylenchus* nematodes overshoot in the soil without creating a cyst, so that all developmental stages are in direct contact with the glucosinolates released during the plant decomposition after its incorporation into the soil. In the experiment of Valdes et al. (2011), eggs and juveniles of *G. rostochiensis* were covered by their cyst walls, which provided a physical barrier for the all compounds that developed during the breakdown of the plants. Lord et al. (2011) inoculated the soil with encysted eggs of *Globodera pallida* (Stone) Berhens and green manure made of three *B.rassica juncea* lines (Nemfix, Fumus, and ISCI99) and the results showed that those containing high concentrations of 2-propenyl glucosinolate were the most effective to induce mortality of the eggs. In addition cruciferous plants were found to be effective to control root-knot nematodes (D'Addabbo et al., 2004).

Besides Brassicaceae, other plants have been tested to suppress plant parasitic nematodes in the soil, like marigolds (*Tagetes*). Among them, the most often used are *T. minuta, T. erecta* and *T. patula* (Wang et al., 2007). The mode of action of marigolds against plant parasitic nematodes is through a biochemical interaction known as allelopathy. Allelopathy is a biological phenomenon by which plants produces one or more compounds that are toxic to other plants, microorganisms or organisms influencing their growth, survival and reproduction. Marigold plants produces many potentially bioactive compounds, among which α-therthienyl has shown nematicidal, insecticidal, fungicidal, antiviral and cytotoxic activities (Wang et al., 2007). For examples, root-knot nematodes *M.eloidogyne arenaria, M. hapla, M. incognita* and *M. javanica* have been successfully suppressed by marigold cultivars of *T. patula, T.erecta, T. signata,* and a *Tagetes* hybrid. Green manure with each one of these plants reduced galling and numbers of second-stage juveniles in subsequent tomato crops compared to the control (Ploeg, 1999). According to Ogden (1997) hypothesis, nematodes are attracted to *T. patula* roots but when they invade them, the roots release ozone, killing the nematodes.

R. graveolens (Rutaceae family) plants used as a green manure against *M. incognita* significantly increased tomato yield and reduced root gall index on tomato roots and soil nematode population density in comparison to untreated control (Sasanelli et al., 2006). The incorporation of a rye cover crop before planting cotton also reduced damage caused by *Meloidogyne M. incognita* (McSorley & Dickson, 1995).

Conclusion

Natural products could provide a sustainable solution to the environmental problems caused by the use of synthetic pesticides and many researchers are trying to identify effective natural products to replace them. Considering the probability of plant secondary products being involved in plant–pest interactions, the strategy of randomly isolating, identifying, and bioassaying these compounds may be an effective method of pesticide discovery. This literature review shows that the use of many herbal extracts, powders or essential oils and their volatile compounds can represent promising valuable methods for the bio-control of plant parasitic nematodes, mainly root-knot nematode species as well as pine wood nematodes. Therefore, the use of plants as natural sources of compounds for the management of plant parasitic nematodes can contribute to minimize the risks and hazards of artificial toxic nematicides, especially for the vegetables produced for fresh consumption, but also for forest tree management, which is particularly problematic in some protected areas.

Acknowledgments

This work was partially supported by the Scientific Grant Agency VEGA of the Ministry of Education and the Slovak Academy of Sciences (Project no. 2/0079/13) (0.5) and by the EU project Research and Educational Centre for Parasitology of the Slovak Academy of Sciences, Operating Program Education (ITMS 26110230045) (0.5).

References

Adegbite, A.A. & Adesiyan, S.O. (2005). Root extracts of plants to control root-knot nematode on edible soybean. *World Journal of Agricultural Sciences*, *1*, 18-21.

Agbenin, N.O., Emechebe, A.M. & Marley, P.S. (2004). Evaluation of neem seed powder for *Fusarium wilt* and *Meloidogyne* control on tomato. *Archives of Phytopathology and Plant Protection*, *37*, 319 – 326.

Al-Banna, L., Darwish, R.M. & Aburjai, T. (2003).Effect of plant extracts and essential oils on root-knot nematodes. *Phytopathologia Mediterranea*, *42*, 123–128.

Alejo, J.C, Suárez, J.M.T., Catzin, S.M, Mendoza, M.N., Baizabal, L.M., Polanco, P.S., Sánchez, S.R.P. & Angulo, M.M.G. (2006). *In vitro* sensitivity of *Meloidogyne incognita* to extracts from native Yucatanean plants. *Nematropica*, 36, 89-97.

Andrés, M.F., Gonzáles-Coloma, A., Sanz, J., Burillo, J. & Sainz, P. (2012). Nematicidal activity of essential oils: a review. *Phytochemistry Reviews*, *11*, 371-390.

Argentieri, M.P., D'Addabbo, T., Tava, A., Agostinelli, A., Jurzysta, M. & Avato, P. (2012). Evaluation of nematicidal properties of saponins from Medicago spp. *European Journal of Plant Pathology*, *120*, 189-197.

Bai, C.Q., Liu, Z.L. & Liu, Q.Z. (2011). Nematicidal constituents from the essential oil of *Chenopodium Ambrosioides* aerial parts. *E-Journal of Chemistry*, *8*,143-148.

Barbosa, P., Lima, A. S., Vieira, P., Dias, L. S., Tinoco, M. T., Barroso, J. G., Pedro, L. G., Figueiredo, A. C. & Mota, M. (2010). Nematicidal activity of essential oils and volatiles derived from Portuguese aromatic flora against the pinewood nematode, *Bursaphelenchus xylophilus*. *Journal of Nematology*, *42*, 8–16.

Begum, S., Qamar Zehra, S., Shaheen Siddiqui, B., Fayyaz, S. & Ramzan, M. (2008). Pentacyclic Triterpenoids from the Aerial Parts of *Lantana camara* and Their Nematicidal Activity. *Chemistry & Biodiversity*, *5*, 1856-1866.

Bello, L.Y., Chindo, P.S., Marley, P.S. & Alegbejo, M.D. (2006). Effects of some plant extracts on larval hatch of the root-knot nematode, *Meloidogyne incognita*. *Archives of Phytopathology and Plant Protection*, *39*, 253 – 257.

Bhattacharya, C., Dasgupta, M.K. & Mukherjee, B. (2012). Management of the root-knot nematode *Meloidogyne incognita* in tea by the two plant extracts, in Tripura, India. *Nematologia Mediterranea*, *40*, 195-201.

Brown, D.J.F., Jones, A.T., Taylor, C.E., Halbrendt, J.M. & Lamberti, F. (1994). An appraisal of some aspects of the ecology of nematode vectors of plant viruses. *Nematologia Mediterranea*, *22*, 253-263.

Caboni P., Aissani N., Cabras T., Falqui A., Marotta R., Liori B., Ntalli N., Sarais G., Sasanelli N. & Tocco G. (2013). Potent nematicidal activity of phthalaldehyde, salicylaldehyde, and cinnamic aldehyde against

Meloidogyne incognita. Journal of Agriculture and Food Chemistry., A-J. DOI: dx.doi.org/10.1021/jf305164m I.

Caboni, P., Sarais, G., Aissani, N., Tocco, G., Sasanelli, N., Liori, B., Carta, A. & Angioni, A. (2012). Nematicidal activity of 2-Thiophenecarboxaaldehyde and Methylisothiocyanate from Caper (*Capparis spinosa*). against *Meloidogyne incognita. Journal of Agriculture and Food Chemistry*, *60*, 7345-7351.

Cavoski, I., Caboni, P. & Miano, T. (2011). Natural pesticides and future perspectives. In: *Pesticides in the Modern World - Pesticides Use and Management*, Stoytcheva, M. (Ed.)., 169-190. ISBN 978-953-307-459-7.

Cavoski, I., Chami, Z.A., Bouzebboudja, F., Sasanelli, N., Simeone, V., Mondelli, D., Miano, T., Sarais, G., Ntalli, N.G. & Caboni, P. (2012). *Melia azedarach* controls *Meloidogyne incognita* and triggers plants defence mechanisms on cucumber. *Crop Protection*, *35*, 85-90.

Chitwood, D.J. (2002). Phytochemical based strategies for nematode control. *Annual Review of Phytopathology*, *40*, 221-249.

Chen, S., Dickson, D.W. & Hewlett, T.E. (1997). Tannic acid effects on hatching of *Heterodera gGlycines in vitro*. *Journal of Nematology*, *29*, 742-745.

Choi, I. H., Park, J.Y., Shin, S. C. & Park, I. K. (2006). Nematicidal activity of medical plant extracts and two cinnamates isolates from *Kaempferia galangal* L. (Proh Hom) against the pinewood nematode (*Bursaphelenchus xylophilus*). *Nematology*,*8*, 359-365.

Choi I.H., Kim, J. Shin, S.C. & Park, I.K. (2007a). Nematicidal activity of medical plant essential oils against the pinewood nematode (*Bursaphelenchus xylophilus*). .*Applied Entomology and Zoology*, *42*, 397-401.

Choi, I.H, Kim, J. Shin, S.C. & Park, I.K. (2007b).Nematicidal activity of onion (*Allium cepa*) oil and its components against the pinewood nematode (*Bursaphelenchus xylophilus*). *Nematology*, *9*, 231-235.

Costa, R., Santos, A. & Ryan M.F. (2003). Effect of *Artemisia vulgaris* rhizome extracts on hatching, mortality, and plant infectivity of *Meloidogyne megadora. Journal of Nematology*, *35*, 437–442.

D'Addabbo, T., Carbonara, T., Leonetti, P., Radicci,V., Tava, A. & Avato, P. (2012). ontrol of plant parasitic nematodes with active saponins and biomass from *Medicago sativa*. *Phytochemistry Reviews*, *10*, 503-519.

D'Addabbo, T., De Mastro, G., Sasanelli, N., Di Stefano, A. & Omidbaigi, R. (2004). Suppressive action of different cruciferous crops on the root-knot nematode *Meloidogyne incognita*. *Agroindustria*, *3*, 379-380.

Decraemer, W. & Hunt, D.J. (2006). Structure and classification. In: *Plant nematology*, Perry, R.N., Moens, M. (Eds). CABI Publishing, Wallingford, 3–32.

Egunjobi, O.A. & Afolami, S.O. (1976). Effect of neem (*Azadirachta indica*). leaf extracts on population of *Pratylenchus brachyurus* on the growth and yield of maize. *Nematologica*, *22*, 125 – 132.

Elbadri, G.A., Lee, D.W., Park, J.C., Yu, W.B., Choo, H.Y., Lee, S.M. & Lim, T.H. (2008a). Nematocidal screening of essential oils and herbal extracts against *Bursaphelenchus xylophilus*. *Plant Pathology Journal*, *24*, 178–182.

Elbadri, G.A., Lee D.W., Park J.C., Yu W.B.& Choo H.Y. (2008b). Evaluation of various plant extracts for their nematicidal efficacies against juveniles of *Meloidogyne incognita*. *Journal of Asia-Pacific Entomology*, *11*, 99–102.

Feeny, P. (1976). Plant apparency and chemical defence.*Recent Advances in Phytochemistry*, 10, 1 – 40.

Gommers, F.J. (1981). Biochemical interactions between nematodes and plants and their relevance to control. *Helminthological Abstract Series B, Plant Nematology*, *50*, 9-24.

Grainge, M.& Ahmed, S. (1988). *Handbook of Plants with Pest-Control Properties*. 1th Edition, New York, Wiley J. & Sons (Eds.) 470 pp.

Grewal, P.S. (1989). Nematicidal effect of some plant-extracts to *Aphelenchoides composticola* (Nematoda) infesting mushrooms, *Agaricus bisporus*. *Revue Nematology*, *12*, 317-322.

Handoo, Z.A. (1998). Plant-parasitic nematodes. http://www.ars.usda.gov/ Services/docs.htm.

Hoste, H., Jackson, F., Athanasiadou, S., Thamsborg, S. M. & Hoskin, O. S. (2006). The effects of tannin-rich plants on parasitic nematodes in ruminants.*Trends in Parasitology*, *22*, 253 – 261.

Ibrahim, S.K., Trabousli, A.F. & El-Haj, S. (2006). Effect of essential oils and plant extracts on hatching, migration and mortality of *Meloidogyne incognita*. *Phytopathologia Mediterranea*, *45*, 238–246.

Ingham, E. (1998). The soil biology primer, soil nematodes. USDA, Natural Resources Conservation Service, Soil Quality Institute.

Isman, M.B. (1999). Pesticides based on plant essential oils. *Pesticides Outlook*, *10*, 68–72.

Isman, M.B. (2006). Botanical insecticides, deterrents, and repellents in modern agriculture and an increasingly regulated world. *Annual Review of Entomology*, *51*, 45-66.

Jain, R.K., Dabur, K.R. & Gupta, D.S. (1994). Assessment of available losses in field due to root knot nematodes *Meloidogyne incognita* in a fewer vegetable crops. *Indian Journal of Nematology*, *24*, 81–184.

Jasy, T. & Koshy, P.K. (1992). Effect of certain leaf extracts and leaves of *Glyricidia aculata* (H.D and K) Steud. as green manure on *Radopholus similis*. *Indian Journal of Nematology*, *22*, 117-121.

Kim J., Seo S.M., Lee S.G., Shin S.C. & Park I.K. (2008). Nematicidal activity of plant essential oils and components from coriander (*Coriandrum sativum*), oriental sweetgum (*Liquidambar orientalis*), and valerian (*Valeriana wallichii*) essential oils against pinewood nematode (*Bursaphelenchus xylophilus*). *Journal of Agriculture and Food Chemistry*, *56*, 7316–7320.

Kim, J., Seo, S. & Park, I.K. (2011). Nematicidal activity of plant essential oils and components from *Gaultheria fragrantissima* and *Zanthoxylum alatum* against the pine wood nematode, *Bursaphelenchus xylophilus*. *Nematology*, *13*, 87-93.

Kirkegaard, J.A., Gardner, P.A., Desmarchelier, J.M. & Angus, J.F. (1993). Biofumigation using *Brassica* species to control pests and diseases in horticulture and agriculture. In: Wratten, N. and Mailer, R.J. (Eds).*9th Australian Research Assembly on Brassicas.* Wagga Wagga, NSW, Agricultural Research Institute, 77-82.

Korayem, A. M., Hasabo, S. A. & Ameen, H. H. (1993). Effects and mode of action of some plant extracts on certain plant-parasitic nematodes. *Anzeiger für Schadlingkunde, Pflanzenschutz Umweltschutz*, 66, 32–36.

Koul, O., Walia, S. & Dhaliwal, G.S. (2008). Essential oils as green pesticides: potential and constraints. *Biopesticide International*,4, 63–84.

Kowalska, B. & Smolinska, U. (2001). Effect of *Brassicaeae* plants on the nematodes *Paratylenchus* spp. and soil microorganisms. *Vegetable Crops Research Bulletin*, *55*, 121-129.

Kutchan T. & Dixon R.A. (2005). Secondary metabolism: nature, chemical reservoir under deconvolution. *Current Opinion in Plant Biology*, *8*, 227- 229.

Lambert, K. & Bekal, S. (2002). Introduction to plant-parasitic nematodes. *The Plant Health Instructor*. DOI: 10.1094/PHI-I-2002-1218-01.

Lamberti, F., Sasanelli, N., D'Addabbo, T., D'Aloisio, V. & De Cosmis, P. (2001). Chemical treatments and soil solarization for the control of the stem nematode, *Ditylenchus dipsaci*, on onions. *Nematologia Mediterranea*, *29*, 149-152.

Lazzeri, L., Tacconi, R. & Palmieri, S. (1993). In vitro activity of some glucosinolates and their reaction products toward a population of the nematode *Heterodera schachtii*. *Journal of Agricultural and Food Chemistry*, *41*, 825–829.

Leela, N. K., Khan, R. M., Reddy, P. P. & Nidiry, E. S. J. (1992). Nematicidal activity of essential oil of *Pelargonium graveolens* against the root-knot nematode *Meloidogyne incognita*. *Nematologia Mediterranea*, *20*, 57- 58.

Li, H.Q., Liu, Q.Z., Liu, Z.L., Du, S.S. & Deng, Z.W. (2013). Chemical composition and nematicidal activity of essential oil of *Agastacherugosa* against *Meloidogyne incognita*. *Molecules*, *18*, 4170-4180.

Li, H.T., Zhao, N.N., Yang, K., Liu, Z.L. & Wang, Q. (2012). Chemical composition and toxicities of the essential oil derived from *Hyssopus cuspidatus* flowering aerial parts against *Sitophilus zeamais* and *Heterodera avenae*. *Journal of Medicinal Plants Research*, *7*, 343-348.

Li. W., Sun, Y.N., Yan X.T., Yang S.Y., Lee S.J., Byum H.J., Moon C.S., Han B.S., & Kim Y.H. et al., (2013). Isolation of nematicidal triterpenoid saponins from *Pulsatilla koreana* root and their activities against *Meloidogyne incognita*. *Molecules*, *18*, 5306-5316.

Lord, J.S., Lazzeri, L., Atkinson, H.J. & Urwin, P.E. (2011). Biofumigation for control of pale potato cyst nematodes: activity of brassica leaf extracts and green manures on *Globodera pallida in vitro* and in soil. *Journal of Agriculture Food Chemistry*, *59*, 7882-90.

Maistrello, L., Vaccari, G. & Sasanelli, N. (2010). Effect of chestnut tannins on the root-knot nematode *Meloidogyne javanica*. *Helminthologia*, *47*, 48-57.

Maistrello L., Vaccari G. & Sasanelli N. (2013). Nematicidal effect of chestnut tannin solutions on the carrot cyst nematode *Heterodera carotae* Jones.Future IPM in Europe. 19-21 March 2013. Book of Abstracts. 166.

McSorley, R. & Dickson, D. W. (1995). Effect of tropical rotation crops on *Meloidogyne incognita* and other plant-parasitic nematodes. Journal of Nematology, *27*, 535-544.

Mian, I.H. & Rodríguez-Kábana, R. (1982). Organic amendments with high tannnin and phenolic contents for control of *Meloidogyne arenaria* in infested soil. *Nematropica*, *12*, 221 – 234.

Miller, P.M., Turner, M.C. & Tomlinson, H. (1973). Toxicity of leaf and stem extract on *Tylenchorynchus dubius*. *Journal of Nematology*,*5*, 173 – 177.

Moosavi, M.R. (2012). Nematicidal effect of some herbal powders and their aqueous exctracts against *Meloidogyne javanica*. *Nematropica*, *42*, 48-56.

Mukhtar, T., Kayani, M.Z.& Hussain, M.H. (2013). Nematicidal activities of *Cannabis sativa* L. and *Zanthoxylum alatum* Roxb. against *Meloidogyne incognita*. *Industrial Crops and Products*, *42*, 447–453.

Natarajan, N., Cork, A., Boomathi, N., Pandi, R., Velavan, S. & Dhakshnamoorthy, G. (2006). Cold aqueous extracts of African marigold, *Tagetes erecta* for control tomato root knot nematode *Meloidogyne incognita*. *Crop Protection*, *25*, 1210-2013.

Nerio, L.S., Olivero-Verbel, J. & Stashenko, E. (2010). Repellent activity of essential oils: A review. *Bioresource Technology*, *101*, 372-378.

Ntalli, N.G.& Caboni P. (2012). Botanical Nematicides in the mediterranean basin. *Phytochemistry Reviews*, *11*, 351-359.

Ntalli, N.G. & Menkissoglu-Spiroudi, U. (2011). Pesticides of botanical origin: a promising tool in plant protection. In: *Pesticides - Formulations, Effects, Fate*, Stoytcheva, M. (Ed) 1-23. ISBN: 9789533075327.

Ogden, S. (1997). Marigolds bite back. *National Gardening,* March-April, 21.

Oka, Y., Nacar, S., Putievsky, E., Ravid, U., Yaniv, Z. & Spiegel, Y. (2000). Nematicidal activity of essential oils and their components against the root-knot nematodes. *Phytopathology*, *90*, 710-715.

Oka, Y., Ben, D., Bat, H. & Cohen, Y (2001). Nematicidal activity of powder and extracts of *Inula viscosa*.*Nematology*, *3*, 735-742.

Olabiyi, T.I. (2008). Pathogenicity study and nematoxic properties of some plant extracts on thre root-knot nematode pest of tomato, *Lycopersicon esculentum* (L.) Mill. *Plant Pathology Journal*, *7*, 45-49.

Padhi, N.N. & Behera, G. (2000). Evaluation of nematicidal potential in ten indigenous plant species against *Meloidogyne incognita*. *Indian Phytopathology*, *53*, 28 – 31.

Pandey, R., Kalra, A., Tandon, S., Mehrotra N., Singh, H.N. & Kumar, S. (2000). Essential oils as potent source of nematicidal compounds. *Journal of Phytopathology*, *148*, 501–502.

Park, I.K., Kim, J., Lee, S.G. & Shin, S.C. (2007). Nematicidal activity of plant essential oils and components from Ajowan (*Trachyspermum ammi*), Allspice (*Pimenta dioica*) and Litsea (*Litsea cubeba*) against pine wood nematode (*Bursaphelenchus xylophilus*) *Journal of Nematology*, 39, 275–279.

Park, I.K., Park, J.Y., Kim, K.H., Choi, K.S., Choi, I.H., Kim, C.S. & Shin, S.C. (2005). Nematicidal activity of plant essential oils and components from garlic (*Allium sativum*) and cinnamon (*Cinnamonum verum*) oils against the pine wood nematode (*Bursaphelenchus xylophilus*) *Nematology*, *7*, 767-774.

Pavaraj, M., Bakavathiappan, G. & Baskaran, S. (2012). Evaluation of some plant extracts for their nematicidal properties against root-knot nematode, *Meloidogyne incognita*. *Journal of Biopesticides*, *5*, 106-110.

Ploeg, A. (1999).Greenhouse studies on the effect of marigolds (*Tagetes* spp.) on four *Meloidogyne* species. *Journal of Nematology*, *31*, 62–69.

Ploeg, A. (2008). Biofumigation to manage plant-parasitic nematodes. In: Ciancio, A. and Mukerji, K.G. (Eds) *Integrated management and biocontrol of vegetable and grain crops nematodes*. Dordrecht, The Netherlands, Springer, 239- 248.

Ray, D., Prasad, D. & Singh, R. P. (2000). Chemical examination and antinemic activity of marigold (*Tagetes erecta* L.) flower. *Annals of Plant Protection Sciences*, *8*, 212-217.

Renčo M. & Sasanelli N. (2013). *In vitro* and *in vivo* effect of chestnut tannin solutions on the potato cyst nematode *Globodera pallida* patotype Pa2. Proceedings of "*10th International symposium ansd schoolon nematology for young scientist*" July 2013, Golytsino, Russia, 131-134.

Renčo M., Sasanelli N., Papajova I. & Maistrello L. (2012). Nematicidal effect of chestnut tannin solutions on the potato cyst nematode *Globodera rostochiensis* (Woll.). Behrens.*Helminthologia*, *49*, 108-114.

Rosa, E., Heaney, R.K., Fenwich, G.R. & Portas, C.A. (1997). Glucosinolates in crop plants. *Horticultural Reviews*, *19*, 99-215.

Ryan, M.F. & Byrne, O. (1988). Plant-insect coevolution and inhibition of acetylcholinesterase.*Journal Chemical Ecology*,*14*, 1965–1975.

Sangwan, N. K., Verma, K. K., Verma, B. S., Malik, M. S., Dhindsa & K. S. (1985). Nematicidal activity of essential oils of *Cymbopogon* grasses. *Nematologica*,*31*, 93-99.

Sarker, S.D., Latif, Z. & Gray, A.I. (2005). Natural product isolation. In: *Natural product isolation*. 2nd Ed. (Sarker S.D., Latifi Z. I., Gray A. Eds). Press, Totowa, New Jersey, 1-25.

Sampson, B.J., Tabanca, N., Kirimer, N., Demirci, B., Baser, K.H.C.B., Khan, I.A., Spiers, J.M. & Wedge, D.E. (2005). Insecticidal activity of 23 essential oils and their major compounds against adult *Lipaphis pseudobrassicae* (Davis) (Aphididae: Homoptera) *Pest Management Science*, *61*, 1122–1128.

Sasanelli, N. (1992). Nematicidal activity of aqueous extracts from leaves of *Ruta graveolens* on *Xiphinema index*. *Nematologia Mediterranea*, *20*, 53-55.

Sasanelli, N., Anton, A., D'Addabbo, T. & Takacs, T. (2006). Nematicidal effect of *Ruta graveolens* L. in horticultural crop protection against root-knot nematodes (*Meloidogyne* spp.) Incontri Fitoiatrici, Turin 2-3 march: 112-114 (In Italian).

Sasanelli, N., Attila, A,.D'Addabbo, T. & Takacs, T. (2007). Nematicidal properties of leaf extracts of *Ruta graveolens* inoculated with arbuscular mycorrhizal fungi. *Russian Journal of Nematology*, *15*, 65-73.

Sasanelli, N., Anton, A., Takacs, T., D'Addabbo, T., Biro, I. & Malov, X. (2009). Influence of arbuscular mychorrizal fungi on the nematicidal properties of leaf extracts of *Thymusvulgaris* L. *Helminthologia*, *46*, 230-240.

Sasanelli, N., Basile, M. & Lamberti, F. (1997). Soil solarization for the control of *Ditylenchus dipsaci* on onion in Southern Italy. Second International Conference on Soil Solarization and Integrated Management of Soilborne Pests. March 16-21. ICARDA, Aleppo, Syria.

Sasanelli, N. & D´Addabbo, T. (1992). The effect of *Cineraria maritime, Ruta graveolens* and *Tagetes erecta* extracts on the hatching of *Heterodera schachtii*. *Nematologia Mediterranea*, *20*, 49-51.

Sasanelli, N. & D'Addabbo, T. (1993a). Potential application of *Ruta graveolens* for controlling *Meloidogyne javanica* on sunflower. *Russian Journal of Nematology*, *1*, 117-120.

Sasanelli, N.& D´Addabbo, T. (1993b). Effect of *Cineraria maritime, Ruta graveolens* and *Tagetes erecta* leaf and root extracts on Italian populations of *Meloidogyne* species. *Nematologia Mediterranea*, *21*, 24-25.

Sasanelli, N. & Di Vito, M. (1991).The effect of *Tagetes* spp. on the hatching of an Italian population of *Globodera rostochiensis*. *Nematologia Mediterranea*, *19*, 135-137.

Handoo, Z.A. (1998). Plant-parasitic nematodes.

Scalbert, A. (1991). Antimicrobial properties of tannins. *Phytochemistry*, *30*, 3875 – 3883

Shoemaker, C.H. & Been, T.H. (1999). Compound models describing the relationship between dosage of (Z)- and (E)-1,3- dichloropropene and hatching of *Globodera rostochiensis*. *Nematology*, *1*, 19–29.

Sittig, M. (1985). Handbook of Toxic and Hazardous Chemicals and Carcinogens. 2nd ed. Noyes Publications, Park Ridge, NJ.

Soler-Serratosa, A., Kokalis-Burelle, N., Rodriguez-Kábana, R., Weaver, C.F. & King, P.S. (1996). Allelochemicals for control of plant-parasitic nematodes. 1. *In vivo* nematicidal efficacy of thymol and thymol/benazldehyde combination. *Nematropica*, *26*, 57-71

Stamp, N. (2003). Out of the quagmire of plant defense hypotheses". *The Quarterly Review of Biology*, *78*, 23–55.

Wiratno, W., Taniwiryono, D., Berg, H.V., Riksen, J.A.G., Rietjens, I.M.C.M., Djiwanti,S.R., Kammenga, J.E. & Murk, A.J. (2009). Nematicidal activity of plant extracts against the root-knot nematode, *Meloidogyne incognita. The Open Natural Products Journal*, *2*,77-85.

Taylor, C.E. & Murant, A.F. (1966). Nematicidal activity of aqueous extracts from raspberry canes and roots. *Nematologica*, *12*, 488-494.

Thoden, T.C., Boppré, M. & Hallmann, J. (2007). Pyrrolizidine alkaloids of *Chromolaenaodorata* act as nematicidal agents and reduce infection of lettuce roots by *Meloidogyneincognita*. *Nematology*, *9*, 343-349.

Thoden, T.C., Boppré, M. & Hallmann, J. (2009). Effects of pyrrolizidine alkaloids on the performance of plant-parasitic and free-living nematodes.*Pest Management Science*, *65*, 823-830.

Traboulsi, A.F., Taoubi, K., El-Hajj, S., Bessiere, J.M. & Rammal, S. (2002). Insecticidal properties of essential plant oils against mosquito *Culex pipiens molestus* (Diptera: Culicidae). *Pest Management Science*, *58*, 491–495.

Valdes, Y., Viaene, N., Perry, R.N. & Moens, M. (2011). Effect of the green manures *Sinapis alba*, *Brassica napus* and *Raphanus sativus* on hatching of *Globodera rostochiensis*. *Nematology*, *13*, 965-975.

Waghorn, G.C. & McNabb, W.C. (2003). Consequences of plant phenolic compounds for productivity and health of ruminants. *Proceedings of the Nutrition Society*, *62*, 383 – 392.

Wang, K.H., Hooks, C.R. & Ploeg, A. (2007). Protecting crops from nematode pests: Using marigold as an alternative to chemical nematicides. Honolulu (HI): University of Hawaii. 6 p. (Plant Disease; PD-35).

Wang, K., Luo, C., Liu, H., Xu, J., Sun, W. & Zhou, W. (2012). Nematicidal activity of the alkaloids from *Macleaya cordata* against certain nematodes. *African Journal of Agricultural Research*, *7*, 5925-5929.

Waterman, P.G. (1999). The tannins – an overview. In: *Tannins in livestock and human nutrition*. Brooker J.D. (Ed). Adelaide, Australian Centre for International Agricultural Research, Australia, 10 – 13.

Zasada, I. A., Meyer, S. L. F., Halbrendt, J. M. & Rice, C. (2005).Activity of hydroxamic acids from *Secale cereale* against the plantparasitic nematodes *Meloidogyne incognita* and *Xiphinema americanum*. *Phytopathology*, *95*, 1116-1121.

Zouhar, M., Douda, O., Lhotský, D. & Pavela, R. (2009). Effect of plant essential oils on mortality of the stem nematode (*Ditylenchus dipsaci*) *Plant Protection Science*, *45*, 66-73.

INDEX

C

D

E

F

G

H

I

N

O

P

Q

T